D1087340

# POTTERY SCIENCE
## Materials, Process, and Products

## ELLIS HORWOOD SERIES IN APPLIED SCIENCE AND INDUSTRIAL TECHNOLOGY

*Series Editor:* Dr. D. H. SHARP, OBE, lately General Secretary, Society of Chemical Industry; formerly General Secretary, Insitution of Chemical Engineers; and former Technical Director, Confederation of British Industry

*Published and in active publication*

**PRACTICAL USES OF DIAMONDS**
A. BAKON, Research Centre of Geological Technique, Warsaw, and A. SZYMANSKI, Institute of Electronic Materials Technology, Warsaw
**PREPARATION, PROPERTIES AND INDUSTRIAL APPLICATIONS OF ORGANOFLUORINE COMPOUNDS**
Editor: R. E. BANKS, Department of Chemistry, University of Manchester, Institute of Science and Technology
**POTTERY SCIENCE: Materials, Processes and Products**
A. DINSDALE, lately Director of Research, British Ceramic Research Association
**MATCHMAKING: Science, Technology and Manufacture**
C. A. FINCH, Managing Director, Pentafin Associates, Chemical, Technical and Media Consultants, Stoke Mandeville, and S. RAMACHANDRAN, Senior Consultant, United Nations Industrial Development Organisation for the Match Industry
**PAINT AND SURFACE COATING: Theory and Practice**
**Vol. 1: Colloid, Surface and Polymer Chemistry of Modern Surface Coatings**
**Vol. 2: Physical Properties of Liquid Coating Compositions and of Derived Films**
Editor: R. LAMBOURNE, lately Section Manager, Research Department, ICI plc Paints Division
**CROP PROTECTION CHEMICALS**
B. G. LEVER, Development Manager, ICI plc Plant Protection Division
**COMMERCIAL ASPECTS OF OILS AND FATS**
R. J. F. LEYSEN, Market Manager, American Soybean Association, Belgium
**HANDBOOK OF MATERIALS HANDLING**
Translated by R. G. T. LINDKVIST, MTG, Translation Editor: R. ROBINSON, Editor, Materials Handling News. Technical Editor: G. LUNDESJO, Rolatruc Limited
**FERTILIZER TECHNOLOGY**
G. C. LOWRISON, lately of Fison's Fertilisers Ltd.
**NON-WOVEN BONDED FABRICS**
Editor: J. LUNENSCHLOSS, Institute of Textile Technology of the Rhenish-Westphalian Technical University, Aachen and W. ALBRECHT, Wuppertal
**MICROCOMPUTERS IN THE PROCESS INDUSTRY**
E. R. ROBINSON, Head of Chemical Engineering, North East London Polytechnic
**QUALITY ASSURANCE: The Route to Efficiency and Competitiveness**
L. STEBBING, Technical Director, Bywater Technology Limited
**POLYMERS IN PACKING TECHNOLOGY**
I. STEPEK, Czechoslovakia. Translation Editor: G. E. J. REYNOLDS, Surrey
**REFRACTORIES TECHNOLOGY**
C. STOREY, Consultant, Durham, one time General Manager, Refractories, British Steel Corporation
**POLYSTYRENE AND ITS MODIFICATIONS**
P. SVEC *et al.,* University of Chemical Technology, Prague
**PETROLEUM TECHNOLOGY**
A. H. SWEATMAN, Visiting Professor, Imperial College and G. A. HOGG, Natural Gas Consultant, lately of British Petroleum Company plc
**PERFUMERY TECHNOLOGY 2nd Edition**
F. V. WELLS, Consultant Perfumer and former Editor of Soap, Perfumery and Cosmetics, and M. BILLOT, former Chief Perfumer to Houbigant-Cheramy, Paris, Presidentd'Honneur de la Societe Technique des Parfumeurs de la France
**THE MANUFACTURE OF SOAPS, OTHER DETERGENTS AND GLYCERINE**
E. WOOLLAT, Consultant, lately of Unilever plc

# POTTERY SCIENCE

## Materials, Process, and Products

ALLEN DINSDALE, OBE, MSc, CPhys, FInstP, HonFICeram, FRSA
Formerly Director of Research
British Ceramic Research Association

*With a Foreword by*
SIR RICHARD BAILEY, CBE
Chairman, Royal Doulton

**ELLIS HORWOOD LIMITED**
Publishers · Chichester

Halsted Press: a division of
**JOHN WILEY & SONS**
New York · Chichester · Brisbane · Toronto

First published in 1986 by
**ELLIS HORWOOD LIMITED**
Market Cross House, Cooper Street, Chichester, West Sussex, PO19 1EB, England

*The publisher's colophon is reproduced from James Gillison's drawing of the ancient Market Cross, Chichester.*

**Distributors:**
*Australia and New Zealand:*
Jacaranda-Wiley Ltd., Jacaranda Press,
JOHN WILEY & SONS INC.
GPO Box 859, Brisbane, Queensland 4001, Australia
*Canada:*
JOHN WILEY & SONS CANADA LIMITED
22 Worcester Road, Rexdale, Ontario, Canada
*Europe and Africa:*
JOHN WILEY & SONS LIMITED
Baffins Lane, Chichester, West Sussex, England
*North and South America and the rest of the world:*
Halsted Press: a division of
JOHN WILEY & SONS
605 Third Avenue, New York, NY 10158, USA

**British Library Cataloguing in Publication Data**
Dinsdale, Allen
Pottery science: materials processes and products. —
(Ellis Horwood series in applied science and industrial technology)
1. Ceramics
I. Title
666 TP807

**Library of Congress Card No.** 85–29573

ISBN 0–85312–980–0 (Ellis Horwood Limited)
ISBN 0–470–20276–9 (Halsted Press)

Typeset by Ellis Horwood Limited
Printed in Great Britain by Butler & Tanner Ltd, Frome and London

# Table of Contents

*To Kathleen*

# Foreword

The art of pottery making is arguably man's oldest profession and a constant fascination to those enmeshed in it, yet much of our knowledge about the scientific principles on which modern industrial practice is based has been gained in little more than a lifetime. No one is more qualified to draw together the threads of that knowledge than Allen Dinsdale. His profound understanding of the subject, his ability to communicate with clarity and appeal and his record of achievements in scientific research and education, have equipped him uniquely for what has surely been a labour of love. Having enjoyed his friendship and had the good fortune to work with him in various capacities for more than thirty years, I feel immensely privileged to introduce this long-awaited book, the tangible result of many hours of painstaking research and dedicated scholarship by a man internationally acclaimed as an authority on the subject.

Richard Bailey
July 1985

# Preface

As more books on ceramics become available, the present addition to the literature may be in some need of justification. It is not a comprehensive and detailed account of pottery materials and processes; the need for this has been adequately met elsewhere. Rather it is an attempt to outline the elementary scientific principles underlying the industrial practice. The art of pottery has been successfully practised for several millennia, but the science of ceramics has not yet celebrated its centenary.

It has often seemed to me that two distinct groups of people can be identified in the pottery industry; those whose business it is to try to understand, and those who have to produce, even when formal understanding is not available. It has been a lifelong interest of mine to try to contribute towards the bridging of that gap, and this book in a sense reflects that attitude. What is attempted here is an introduction to the scientific principles that are relevant to an understanding of pottery materials and processes. The approach is essentially that of a physicist, and, in order to ensure that the material is of use to a wide constituency of practising ceramists, the treatment is not more advanced than A level. University or college students should find the text interesting as an introduction to a wider and more advanced course in ceramics.

The scope has been restricted to pottery, but this word in itself requires some definition. Nomenclature in this field is notoriously erratic and ambivalent. For example, the word china may be used in a generic sense to cover a wide range of articles that might be found in a china cabinet, irrespective of what they were made of; or it may equally well be used to describe specifically a particular material, with well-defined characteristics. In the present context, pottery is defined in terms of function; the scope of the book covers tableware, tiles, sanitary ware, and electrical porcelain. In terms of body systems they are mainly based on the triaxial combination clay-flux-quartz, with the addition of the special case of bone china. For our purposes, and with no implications with respect to appropriate nomenclature in other fields, the triaxial system fired to a significant degree of porosity is called earthenware; when fired to a very low porosity, it may be called vitrified ware, china, or porcelain; when containing a substantial proportion of bone it is referred to as bone china. Where the term whiteware is used, it refers not to undecorated ware, but to bodies containing refined white-burning clays, as distinct from unrefined clays giving a buff or red colour.

In order to make the book acceptable internationally, I have used SI units wherever appropriate, but I am conscious of the fact that other units, such as pint weight for slips, are still in common use in industry. Accordingly, I have included a table of conversion factors.

I must place on record my indebtedness to many people who have helped me in a variety of ways. I owe most to the colleagues with whom I had the privilege of working at the British Ceramic Research Association. Over a period of several decades, they made a massive impact on the development of scientific thinking in ceramics, and my debt to them can be judged by the frequency with which their names occur in the references at the end of the chapters. Dr R. C. P. Cubbon and Mr W. Roberts have been kind enough to read the manuscript and make helpful suggestions. I would like to thank Dr David Sharp, who first suggested the project, and my publishers for all the help they have given me.

I have made frequent use of matter appearing in the publications of the British Ceramic Society, and I record with thanks their permission to make use of the information in Table 3.1, and Figs 4.2, 4.3, 4.5, 4.6, 5.5, 5.8, 6.3, 7.3, 10.2, 10.3, 12.1, 12.5, 13.5, 15.2, 15.8, 15.9, 17.3, 18.1 and 18.2. I would also like to thank Dr D. W. F. James, Chief Executive of the British Ceramic Research Association Ltd, for permission to use the data in Figs 7.2, 7.4, 9.5, 9.8, 11.6, 11.7 and 18.3.

The fact that Sir Richard Bailey, who has done so much to advance the interests of the pottery industry, has consented to contribute a Foreword is only the latest of a long line of kindnesses that he has shown to me during almost a lifetime of friendship and collaboration in various spheres, all deeply appreciated.

Finally, I must record my thanks to my wife, Kathleen, and my son, Paul, for constant interest and encouragement; and to my daughter-in-law Christine, for so skilfully and diligently deciphering my handwriting and producing from it a legible typescript.

Allen Dinsdale
1985

# 1

# Introduction

The production of pottery is one of the oldest of man's activities. It is based on the exploitation of the unique properties of clay, namely that it can be moulded into shapes and that these shapes assume a state of permanence on heating. Shaping must have come first, from man's interaction with mud, and the obvious features of footprints, for example. Hardening in the sun would be followed by the discovery that higher temperatures in a wood fire produced even harder material.

The permanent nature of fired pottery has given it a high degree of importance to archaeologists; their finds provide valuable clues to the habits and movements of early settlements. In some cases, sites have been occupied successively by many different societies, leaving broken pottery as clues to the nature of social changes. The long history of pottery is now well documented, but is too detailed to be dealt with in full here. It is, however, worthwhile noting some of the more important developments.

The beginning is uncertain but may well have been as long ago as 15 000 BC; certainly burnt ceramic products have been found in the Nile Valley dating back to around 11 000 BC. A crude form of earthenware, glazed with an alkaline glaze was being made in Egypt and the near East in 5000 BC. Terracotta vases were placed in tombs to provide sustenance for spirits of the departed as early as 4000 BC, and tiles with a blue copper glaze were being made about 3500 BC. The Chinese were making ware by 3000 BC, around which time the potters' wheel was being used, and fired clay was being used for moulds. In the millennium before Christ the Chinese were making hard porcelain, Greek pottery was at its height, with great beauty of form and striking hand-painting of vases. There are references in the Bible to clay, the wheel, and the fire.

The oldest pottery found in England dates back to the first century AD and the Roman occupation. Between then and 1500 AD the most important development was that of transparent porcelain by the Chinese, typified by the pottery of the Sung (960–1280) and Ming (1370–1640) dynasties. Activity in England began again with the Cistercian ware in the early sixteenth century, their ware being crude earthenware jugs with a green or yellow glaze. Pottery was made at Lambeth in the middle of the sixteenth century, and the first porcelain made in Europe was in Florence in 1580. The seventeenth century saw the beginnings of a British pottery industry, with the Toft brothers making slipware in Staffordshire, Dwight making delft ware and salt-glazed soft porcelain at Lambeth, and the Elers brothers exploring the possibilities of washed clays. But the great impetus came in the eighteenth century, which saw the beginnings of developments that have made the industry what it is today. In the first quarter of the century, Astbury had introduced flint and Devon and Dorset ball clays into the earthenware body, and fifty years later Cookworthy had discovered china clay. Thus the way was open for the replacement of red or buff bodies, based on local clay deposits, by a white-burning body. Slip casting in plaster moulds was being investigated, and at the other end of the process, a form of transfer printing was in use. Soft-paste porcelain was first made in England at Chelsea and Bow in 1745–50. Rapidly following Cookworthy's introduction of china clay and Cornish stone, Spode developed the bone china body at the end of the century. The two largest groups of tableware manufacturers today, Doulton and Wedgwood, have their origins in these times of innovation. John Doulton worked for Dwight at the Fulham pottery, and started the Doulton pottery at Lambeth in 1815, eventually making a wide range of ceramic products in addition to tableware. Henry Doulton entered the North Staffordshire scene when he acquired a small pottery in Nile Street, Burslem, in 1877. Josiah Wedgwood entered the industry at the early age of nine in 1739, and over the next fifty years developed the cream earthenware, known as Queensware, black basalt for ornamental ware, and, after many trials, the well known Jasper body. The reason for the concentration of the industry in North Staffordshire was the availability in the area of both clay and coal.

Other branches of the industry began to take shape in the nineteenth century. A growing concern about sanitation was the impetus for developments in sanitary ware based on Harrington's water closet, and by 1850 plumbing fixtures were being made by Lambeth and Thomas Twyford. In the latter part of the century, the introduction of electricity gave rise to the beginnings of the electrical porcelain industry. Examples of modern products are shown in Figs 1.1, 1.2 and 1.3.

The size of the modern pottery industry in the United Kingdom may be judged from the statistics given in Table 1.1, which shows the value of UK products sold in the home market, and of those exported, for the year 1984.

(a)

(b)

Fig. 1.1 – Bone china products. (Courtesy Royal Doulton.) (a) 'Carnation' from the 'Kind of Loving' range. (b) 'Europa and the Bull' from the 'Myths and Maidens' series.

Fig. 1.2 – Wall tiles and sanitary ware in a modern bathroom setting. (Courtesy H and R Johnson Tiles.)

Fig. 1.3 – Electrical porcelain. Large insulator for 400 kV service. manufactured by Allied Insulators Limited for GEC Hixon Works.

These data are in units of £million, and are taken from *Business Monitor.* It will be noted that a substantial proportion of the production is exported, this being particularly significant in the case of tableware, kitchenware, and ornamental ware.

**Table 1.1** – Sales of ceramic products by UK manufacturers, 1984

| | | Home | Export | Total |
|---|---|---|---|---|
| China or porcelain | Tableware and kitchenware | 37.5 | 69.1 | 106.6 |
| | Other | 38.6 | 16.8 | 55.4 |
| Stoneware | | 11.9 | 3.3 | 15.2 |
| Other pottery (including | Tableware and kitchenware | 87.0 | 68.2 | 155.2 |
| earthenware) | Other | 8.8 | 9.2 | 18.0 |
| Glazed tiles | | 74.9 | 9.1 | 84.0 |
| Sanitary ware | | 96.5 | 26.0 | 122.5 |
| Electrical ware | | 29.2 | 21.0 | 50.2 |
| | Total | 384.4 | 222.7 | 607.1 |

The manner in which the pottery industry has adjusted to developments in the physical and chemical sciences makes a fascinating study. When the growth of the industry in England was gaining momentum, in the latter part of the eighteenth and early part of the nineteenth century, there was not much help available from the scientists, for science itself was in a very primitive state by modern standards. Alchemy was still in the field, and Wedgwood was elected to the Royal Society as a result of his work on pyrometry, which was of a fairly elementary nature. Nevertheless, Wedgwood was something of a scientist by instinct, as an examination of his records of experiments will show. He also made contacts with scientists, for example with Priestley, who discovered oxygen in 1774.

But the great advances in science were to come at the end of the nineteenth century, with radioactivity, x-rays, the electron, and the Planck quantum. Willard Gibbs, in 1874, had already enunciated his phase rule, the application of which has done so much to establish the nature of the equilibrium phases in high temperature ceramic systems. The tremendous developments of the twentieth century were to follow, beginning with a revolution in man's understanding of the nature of matter, and leading to his manipulation of it for technological ends. As the ideas emerging from fundamental scientific understanding were progressively translated into technology, they presented ceramists with a very powerful array of new techniques for examining structure and constitution on a smaller and smaller scale. Such include x-ray diffraction, optical and electron microscopy, EPMA, and instrumental chemical analysis.

A parallel development in the pottery industry was the growth of a common conscious need to apply the scientific method to industrial problems. The English Ceramic Society was formed in 1900, following closely on the establishment of a similar body in America. Encouraging exchange of technical information in meetings, and by publication, these bodies have played a crucial part in the development of scientific attitudes in the industry. In England, a major contribution was made by J. W. Mellor, author of *A Comprehensive Treatise on Inorganic and Theoretical Chemistry,* who stimulated both research and teaching over a span of some thirty years. It is interesting to note that the 1923 Transactions of the Ceramic Society record a correspondence between Mellor and Bragg, whose x-ray studies on crystal structure had led him to be very interested in minerals. He carried out some studies on kaolinite, to help identify the crystal phases that developed on heating to 1000 °C. Co-operative research, with Government support, has been prosecuted through the Research Association, whose staff have explored in some depth almost every area of pottery materials, processes and products, where science has any part to play. Teaching and research have also been carried on at Stoke, Sheffield, and Leeds.

Looking at the picture today, we see the application of a wide range of modern techniques to reveal the properties of materials and products in relation to their chemical and physical structure. A new understanding of the common principles applying to a range of materials, including ceramics, glass, metals, and polymers has encouraged the broad concept of a new discipline, rather misleadingly known as Materials Science.

But one needs to ask how far all this knowledge has permeated the industrial scene as far as pottery is concerned. The answer must be not far enough as yet, and there are several reasons for this. The first is that many academic research workers confine their attention to pure materials. There is justification for this, in that basic principles can only be established by reducing the number of variables and simplifying the problem. Nevertheless, this approach in itself limits the applicability of their findings. There is no easy transfer of findings on pure materials to complex minerals in use in industry. Few research workers are willing to follow the alternative course, which is to subject materials in bulk to imposed conditions and to study the response. Modern availability of computerized techniques should make it easier to deal with multivariable systems. A deeper reason, however, is that the properties of pottery materials and products are dominated by texture rather than structure. In technical ceramics, the fine detail of crystal structure, and the influence of defects, are of critical importance in determining the possibility of obtaining improvements in function. But in pottery bodies, their influence is masked by the presence of pores, cracks, and other inhomogeneities. This is not to say that knowledge gained on the frontiers of ceramic research is not of use; but rather to suggest that its use is in terms of general understanding rather than in particular application.

So what does the future hold for pottery in this science-dominated age? Many old skills have already disappeared, and new skills will be needed. New processes, like plate pressing, pressure casting, fast firing, and direct decoration, will replace the old. The areas of likely change in the tableware industry have been suggested in a perceptive paper by Secker *et al.* (1983). There is bound to be a widening gap between industry-based and studio-based pottery production, reflecting respectively the dominant influence of science and art.

But that is beyond the scope of this book. What is attempted here is an outline of the scientific principles involved in pottery manufacture. The approach is mainly physical, which means that consideration of the properties of matter and energy, and their interaction with each other, predominate. The sequence of ideas follows a traditional pattern; raw materials, individually and mixed together; body design; the role of water; forming; drying; firing; glazes and decoration; and finally some of the more significant physical properties of the finished products.

## REFERENCE

Secker, P. E., Barker, P. H., Green, J. T., Nicholas, J. A. and Till, J. R. (1983) *Tr. & J. Brit. Cer. Soc.*, **82**(4), 120.

# 2

# Body materials

Ceramic products can be made from a very wide range of inorganic materials, but the practical choice is severely restricted by the special properties needed in the manufacture of modern high quality goods. The technical requirements are so demanding and numerous that only a limited range of materials can meet them. This chapter is concerned only with those materials that fall into this category, and that are, therefore, in common use. No attempt is made to deal in detail with either the availability or the geological distribution of the various minerals, as this is widely documented elsewhere. Rather the aim is to identify those characteristics that are of crucial importance, and the possession of which makes these materials fit for the purposes for which they are used.

It is convenient, though admittedly arbitrary, to divide pottery body materials into three groups, distinguished by the distinctive physical characteristics that fulfil the functional requirements. In the first class are those materials that assist in the forming process. This may take the form of plastic working, slip casting, extrusion, or dust pressing. In all cases certain basic needs have to be satisfied. The material must deform easily and without rupture. When deformed it must retain the imposed shape, and have sufficient strength in the unfired state to withstand handling and other subsequent processes. These characteristics are present to an optimum degree in clay, so that all pottery produced in bulk on a commercial scale contains a substantial proportion of clay of one kind or another. Some products, indeed, are made wholly from clay, but the great majority of products have other materials in combination with clay.

The second component is usually a material known as flux. The function of the flux is partially to melt, and react with other materials, during the firing, and to produce a glassy matrix, which increases the strength of the fired product.

The remainder of the composition is made up by the inclusion of a relatively non-reactive material. This provides a rigid component, but may also confer some very important physical property, such as thermal expansion.

A common requirement of most whiteware products is that the materials shall be of such a degree of fineness as to make processing easy and to give a smooth finish to the article. The subject of fineness will occur again and again in this book, as its significance can hardly be exaggerated, in so far as many physical properties are markedly influenced by it. Suffice it to note at this point that whereas for many materials fineness has to be achieved by grinding, it is an important feature of clay that it is fine-grained in the natural state.

It is therefore useful to think of a whiteware body system as comprising these three elements: clay, flux and filler, and the most commonly used materials will now be considered under these headings.

## 2.1 CLAY

Ceramic products were among the earliest human artefacts to emerge many thousands of years ago. The process began with the discovery by primitive man that earth mixed with water could be moulded into shapes that could be used, for example, for drinking and washing vessels, or storage vessels for oil and grain. It was also found that when these articles were heated they acquired a measure of strength and permanence. Glazing came much later, and the use of multi-component body systems is a comparatively recent development.

In the UK small manufacturing units grew up in localities where beds of suitable clay were found often in conjunction with available coal. Many traditional products, such as stoneware and red clayware, were made from clay alone. These clays contained many colouring contaminants, and could not be used in the context of the modern whiteware industry, where there is a stringent demand for purity. The emphasis has thus shifted from localized clay-based industry, to a clay-producing industry in its own right.

Britain is exceptionally fortunate in having major deposits of high quality clays in the south-west (Bailey, 1974; Mitchell, 1974). The technology of extraction and beneficiation has advanced at a great rate in recent years, so that clays can now be supplied in dry shredded or powdered form

They can be blended so as to provide tailor-made properties for specific purposes. Alongside these advances in processing, decades of scientific research have produced new understanding of the constitution of clay, and of its technological response. A vast amount of technical data is supplied by the clay producer, but discussion of the wide range of available properties is beyond the scope of this book. So also is the detailed mineralogy, which is now well understood. It should be noted, however, that clay is not a simple substance. In addition to kaolinite, which is the principal mineral, there are generally present a number of other clay minerals, some of which, like montmorillonite, can have an important influence on body properties even when only present in very small proportions.

The two principal types of clay used in the whiteware industry are china clay and ball clay. Well over a quarter of a million tons of these are used annually in the tableware, tile and sanitary industries, in roughly equal proportions. Although these two clays are similar in that they both consist mainly of kaolinite, and although they both come initially from the same source, they differ greatly in their characteristics and are complementary in their contributions to whiteware body technology. The essential difference between them is perhaps best understood by considering their origins.

Fig. 2.1 – Hydraulic extraction of china clay. (Courtesy ECC International)

Both were originally formed by a process of decomposition of the felspar contained in granite rock, such as the formations typically found in Cornwall. Way back in geological time, superheated steam and hot acid gases emerging through fissures in the Earth's crust attacked the felspar; partial decomposition changed the felspar into kaolinite, or china clay, the alkaline components being washed away. This kaolinization process may be represented as follows:

| Felspar | + Water | → | Kaolinite | + | Free silica |
|---|---|---|---|---|---|
| $K_2O.Al_2O_3.6SiO_2$ | $+ 2H_2O$ | → | $Al_2O_3.2SiO_2.2H_2O$ | + | $4SiO_2$ |

The china clay so produced has to be extracted from the surrounding rock by high pressure water jets, and the resulting slurry initially contains a number of other minerals of which quartz and mica are common. These can be removed by subsequent separation techniques, so that the resulting clay supplied to the industry is very high in kaolinite content.

However, at various geological times, some of the kaolinite was washed out and transported by rivers to other areas, where it was deposited as thick layers of what is now known as ball clay, so called because it was originally mined by cutting out large cubes of clay. Large deposits of these transported clays are found in Devon and Dorset. They have two important features that derive from this process of transportation. One is that they can contain substantial quantities of organic matter picked up en-route. This is not necessarily harmful, as it can help with plasticity, and in any case most of it burns out on firing, leaving a white fired product. The other characteristic is a very fine grain size, fine particles being transported and coarse grains being left behind. The question of fineness will be considered in more detail later.

At this point it is necessary to deal with the complex nature of clay, which may contain, in addition to kaolinite, other important minerals such as quartz and mica. Many investigatory techniques are now available for the direct determination of mineral content, but the traditional technique of inferring the mineralogical composition from the chemical analysis is still useful in many circumstances, especially where more primitive clays of doubtful quality are concerned.

In this treatment, a so-called rational analysis is derived by allocating the known oxides listed in the chemical analysis to a range of minerals presumed to be present. An example will serve to show how the calculation in carried out.

Suppose a clay shows the following composition in terms of constituent oxides: $SiO_2$ 51.0%, $Al_2O_3$ 32.0%, $Na_2O$ 0.1%, $K_2O$ 2.5%, remainder 14.4%. The remainder would include the loss on ignition, and some impurities containing such elements as calcium, magnesium, titanium and iron, which we will ignore for present purposes.

Suppose now that the rational analysis comprises the minerals kaolinite, quartz, soda mica (paragonite), and potash mica (muscovite). Assuming whole numbers for the atomic weights, the relevant molecular weights are as follows:

| | | *Molecular weight* |
|---|---|---|
| Kaolinite | $Al_2O_3.2SiO_2.2H_2O$ | 258 |
| Quartz | $SiO_2$ | 60 |
| Soda mica | $Na_2O.3Al_2O_3.6SiO_2.2H_2O$ | 764 |
| Potash mica | $K_2O.3Al_2O_3.6SiO_2.2H_2O$ | 796 |
| | $Al_2O_3$ | 102 |
| | $K_2O$ | 94 |
| | $Na_2O$ | 62 |
| | $H_2O$ | 18 |

From these the percentage oxide concentrations in the four minerals can be calculated from the appropriate oxide molecular weights.

| | $Na_2O$ | $K_2O$ | $Al_2O_3$ | $SiO_2$ | $H_2O$ |
|---|---|---|---|---|---|
| Kaolinite | – | – | 39.54 | 46.51 | 13.95 |
| Quartz | – | – | – | 100.00 | – |
| Soda mica | 8.12 | – | 40.05 | 47.12 | 4.71 |
| Potash mica | – | 11.81 | 38.44 | 45.23 | 4.52 |

The first step in calculating the rational analysis is to use the $Na_2O$ and $K_2O$ percentages in the chemical analysis to determine the mica content. Thus,

$$\text{Soda mica} = \frac{0.1}{8.12} \times 100 = 1.23\%$$

and

$$\text{Potash mica} = \frac{2.5}{11.81} \times 100 = 21.17\%$$

The next step is to consider the alumina content.

Soda mica contributes $1.23 \times 0.40 = 0.49$
Potash mica contributes $21.17 \times 0.38 = 8.13$

So the total contributed by the mica is 8.62% $Al_2O_3$.

This means that the $Al_2O_3$ contributed by the kaolinite is $32.0 - 8.62 = 23.38\%$, and the percentage of kaolinite corresponding to this is thus $23.38/39.54 \times 100 = 59.10\%$.

Finally, the amount of free or uncombined silica can be computed, and this will indicate the presence of quartz.

The $SiO_2$ in combined form is,

| | | |
|---|---|---|
| from the soda mica | 1.23 × 0.471 = | 0.58% |
| from the potash mica | 21.17 × 0.452 = | 9.57% |
| from the kaolinite | 59.10 × 0.465 = | 27.49% |
| Total | = | 37.64% |
| So the free silica is | 51.00 − 37.64 = | 13.36% |

The rational analysis is thus,

| | |
|---|---|
| Kaolinite | 59.1% |
| Quartz | 13.4% |
| Soda mica | 1.2% |
| Potash mica | 21.2% |
| Miscellaneous | 5.1% |

This method of inferring the mineralogical composition of clay is now largely obsolete in industries where sophisticated techniques are available to give more direct information, but where such techniques are not available it is a useful guide for interpreting data obtained by chemical analysis.

Some of the ways in which the properties of clay contribute to the required characteristics of pottery bodies are now considered. Among the more important aspects are particle size, plasticity, and strength.

### 2.1.1 Particle Size

Before dealing in detail with particle size, some comment is necessary on the question of shape. In many areas of ceramic technology, the arguments rest on the assumption that the particles are equant; that they approximate in shape to cubes or spheres. For many substances that are ground to the required degree of fineness this is roughly true, but in the case of clay it is a very large assumption. In so far as clay often flocculates in such a manner that its physical behaviour is related to the clusters, they may be considered as equant in nature, but the unit kaolinite particle is very different.

X-ray analysis shows that kaolinite consists of a layer structure in which planes of Si–O atoms alternate with planes of Al–O or OH atoms, with a unit cell dimension of the order of $10 \times 10^{-10}$ m. A side view of the arrangment of atoms is shown in Fig. 2.2. There are, of course, many hundreds of unit cells in a Kaolinite crystal. This crystal usually occurs in the form of a thin, roughly hexagonal plate, as can be seen from the electron microscope picture shown in Fig. 2.3. These plates have an axis ratio of the order of 10:1. Typically the diameter would be about ½ to 1 $\mu$m, and the thickness 0.05 $\mu$m. They are normally found stacked together in large numbers, with their faces parallel to each other, as may also be seen in Fig. 2.3.

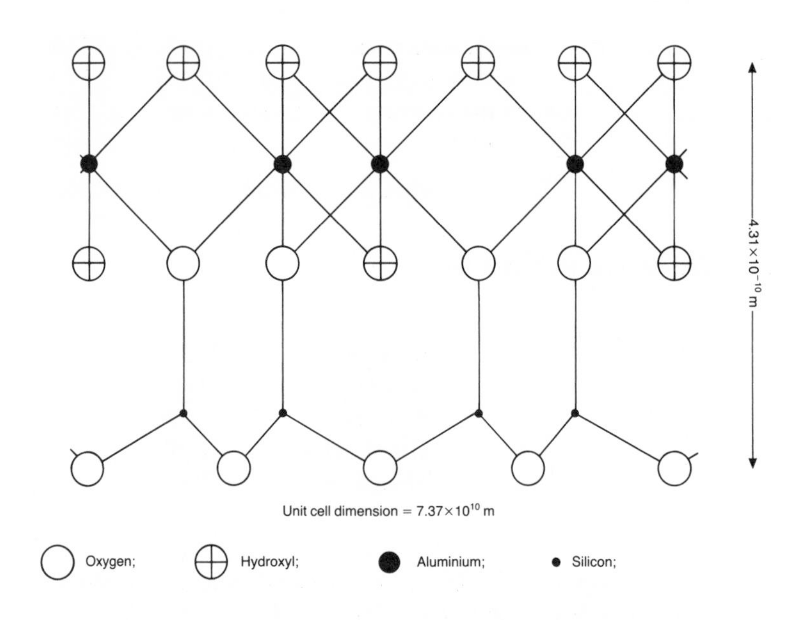


Fig. 2.2 – Layer structure of kaolinite.

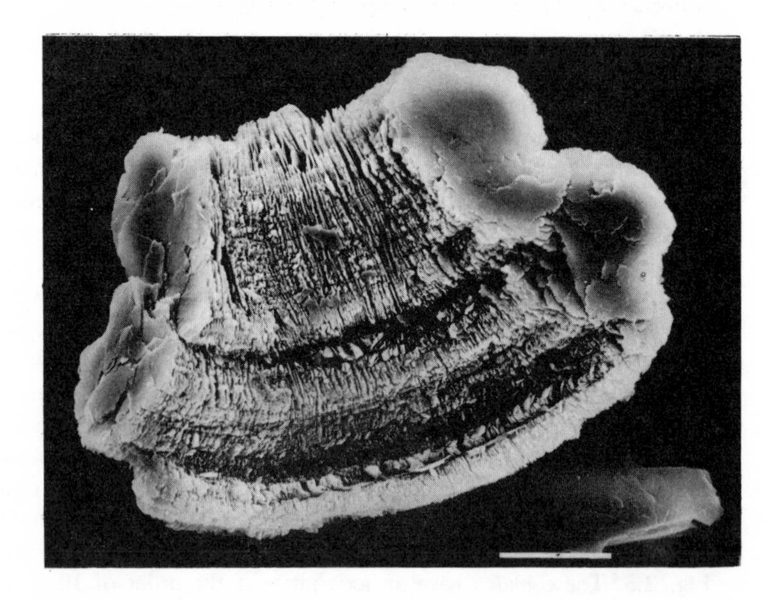

Fig. 2.3 – Stacked kaolinite crystals in china clay. Dimension marker is 10 $\mu$m long (Courtesy ECC International).

Particle size measurements on clays often involve sedimentation techniques, accelerated by the use of the centrifuge, and again the assumption is made that the settling rate is the same as that of an equivalent sphere. Sizes are thus often quoted in terms of an equivalent spherical diameter. Most of the particle sizes quoted in this book are on that basis.

We turn now to look at the measured size distribution of the materials commonly found in pottery bodies. In the classification of sedimented materials, geologists often use the term clay for those sizes that come between colloids on the one hand and silt on the other, and consider them as falling into the range 0.15 to 4 $\mu$m. This is not inconsistent with the values shown in the typical distribution curves shown in Fig. 2.4. These so-called cumulative curves show the weight percentage of the material finer than a specified size. Confining our attention to the clays, we see first of all that the finest fraction of ball clay can contain an appreciable amount of material with a size of 0.05 $\mu$m, which is approximately the size of the crystals of kaolinite mentioned above. For spheres of diameter 0.05 $\mu$m, and specific gravity 2.5, it can be calculated that the specific surface area is of the order of 50 $m^2/g$. Gas adsorption measurements on clays often yield values in the range 10 to 100 $m^2/g$.

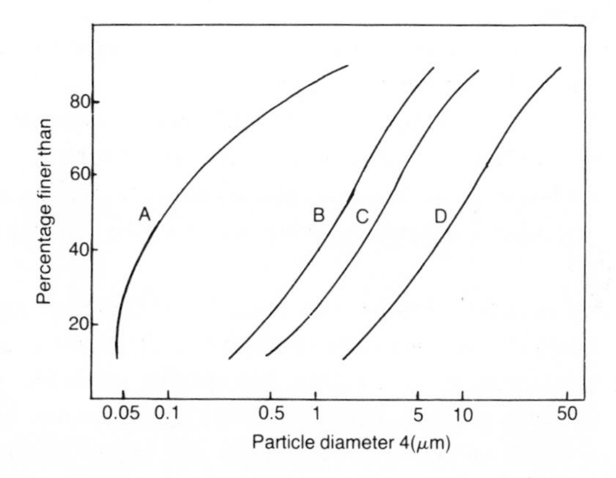


Fig. 2.4 – Typical size distributions for some body materials. A, Ball clay; B, China clay; C, Bone; D, Flint or stone.

The graphs show that ball clay is substantially finer than china clay, and that both are finer than the ground materials with which they are frequently mixed in the compounding of bodies. This degree of fineness enables bodies with high packing density to be made, and it also confers advantages in relation to some of the other technological properties that are of industrial significance. Among these are plasticity and strength.

### 2.1.2 Plasticity

When clay is mixed with water it exhibits a phenomenon popularly known as plasticity, which is easier to describe than it is to define or measure. In scientific terms, the stress-strain relationships are both exceptional and complex. They will be dealt with in more detail in a later chapter; they have long engaged the attention of research workers in the ceramic field, but a full understanding is still some way off.

Some qualitative observations can, however, be usefully made at this point. For a material to be suitable for a plastic-making process, such as the forming of plates or cups, it needs to satisfy the following criteria:

(a) It must be capable of undergoing large strains and large rates of strain without rupture.
(b) It must show no deformation at low levels of stress. In practice, this means that it must be able to retain the shape that has been imposed upon it.

The fine grain size of clay, the plate-like nature of its particles, and its special relationship with water, all contribute to a stress–strain behaviour that enables it to meet these criteria. This, above all else, is the characteristic that makes it an indispensable element in pottery body formulation.

### 2.1.3 Strength

The *Concise Oxford Dictionary* definition of clay contains the word ‘tenacious’; and the definition of that word, in turn, contains the phrase ‘strongly cohesive’. We might expect, then, that clay would make a significant contribution to the strength of bodies of which it was a major constituent.

Three aspects of the strength of pottery bodies are important, namely unfired strength (sometimes called green strength), in-kiln strength, and fired strength.

When an article has been formed, by whatever process, and is subsequently dried, it needs to have adequate strength in order to withstand stresses encountered in fettling, transport, or stacking. Non-plastic materials, such as flint and stone, have very little strength when dried out. On the other hand, clay maintains an adequate level of strength even when the water has been removed. It is not obvious why this should be so. Dry clay powder has very little strength when compacted. If mixed with water and formed into a given shape, it has quite a high strength when dried out. It is possible that the surface tension, owing to the water, brings the particles into very close contact, thus bringing close-range attractive forces into effect. It may also be the fact that the water leaves behind residual soluble matter that acts as a cement. In either case, the fine particle size of clay, resulting in a very large number of points of contact per unit volume, gives rise to sufficient unfired strength to meet the requirements. Ball clay may have an unfired strength as high as $10 \times 10^6$ $\text{Nm}^{-2}$; china clay, because of the coarser grain, more like half this. Mixed with non-plastic materials, body strengths of the order of 2 to $5 \times 10^6$ $\text{Nm}^{-2}$ are achieved; sufficient for most practical purposes.

Plasticity and strength can, of course, be achieved by the use of organic binders instead of clay. It is also possible to use inorganic polymers for the same purpose. It is important to note, however, that for many purposes it is essential to have adequate strength, not only at room temperature but also during the early stages of the firing. This is especially the case when articles, such as plates, are fired in bungs, so that there is a considerable load on the lower members. Organic materials burn out at relatively low temperatures, so that at 300 °C bodies relying on them for strength can become extremely fragile. Clay-based bodies, by contrast, maintain their high level of dry strength to temperatures well beyond this. The kaolinite molecule does not lose its combined water until around 500 °C, and beyond this sufficient strength is maintained until the fired strength begins to develop.

Fired strength, in general terms, is associated with the development of strong crystals, such as mullite, either joined together by, or embedded in, a glassy matrix. Clay serves the dual purpose of entering into both these processes. Heating kaolinite to high temperatures results in the production of mullite, $3Al_2O_3.2SiO_2$, and this contributes markedly to the strength of many fired body systems. On the other hand, when alkalis are present, either in the clay or in other body materials, akali-alumino-silicate glass can be formed, thus providing some of the glassy matrix. In fact, fired clay alone can produce strong vitreous bodies, as in the case of the traditional high temperature stoneware.

On all these grounds, then, it is clear that clay is an essential component for a wide range of industrial pottery products. Of course, not all its properties are favourable, and some of the problems that can stem from its use will be noted later in this book. But its inherent fineness, achieved without any recourse to grinding, its particle shape, its unique plastic properties, and its strength, all combine to give it a prime place among ceramic materials. That it is also relatively easy to free from injurious impurities, fires to an acceptable colour, and is commercially available in very large tonnages, means that it is virtually indispensable.

## 2.2 FLUX

The function of the flux in pottery bodies is to promote the formation of a limited and controlled amount of glass, which plays an important part in cementing the crystalline components together. The relationship of the properties of the glass to those of the crystals will be dealt with in detail in a later chapter, but it can be noted at this point that it is of crucial importance in determining many of the significant technical properties of the body system.

It is true, of course, that many clays are vitrifiable in themselves, but usually need to be fired to a fairly high temperature in order for this to be of practical use. In some body systems, such as the so-called soft porcelains, ground glass itself is added to the body. But in most of the systems with which we are concerned, some additional fluxing material is required. Consideration of modern theories of glass structure would indicate that minerals containing soda or potash would be likely to be suitable, since both these ions are known to be powerful glass formers. If they can be found in conjunction with non-melting elements, so that vitrification can be gradual and easy to control, the problems associated with firing are likely to be diminished. The range of fluxes available has been reviewed by Royle (1974). The felspar minerals meet these requirements, and have, therefore, become the most widely used for this purpose.

Felspar itself occurs in igneous rocks, such as granite, and has a number of crystalline forms. The two varieties of most interest in the whiteware ceramic industry are potash felspar (orthoclase). $K_2O.Al_2O_3.6SiO_2$ and soda felspar (albite). $Na_2O.Al_2O_3.6SiO_2$. Both of these provide an abundant source of insoluble alkali. For a number of technical reasons, potash felspar is generally preferred.

World-wide, felspar is widely used as a flux in porcelain, and has come into use in recent years in bone china. Batchelor (1973) has described the winning of this material by the flotation process. It is not, however, found in quantity in Britain, and the traditional industry here has been built up on a cheaper material, known as Cornish stone or China stone. Large deposits of this material, are found in Cornwall, where it appears as an intermediate stage in the kaolinization of granite. The stone is a heterogeneous mixture of clay, felspar, quartz, unaltered granite and a number of accessory minerals in small proportions. The different grades available contain these components in different proportions, varying from hard purple to white or buff stone. The stone normally contains about 70–75 per cent silica, 12–15 per cent alumina, and about 8 per cent alkali. This compares with an alkali content of 17 per cent in orthoclase, and 12 per cent in albite. Commercial felspar contains about 12 per cent alkali.

This means that the fluxing action of Cornish stone is much less rapid than that of pure felspar, and this has many advantages in practice. The stone is obtained in the form of rock from open quarries, and ground to a grain size distribution such as that shown in Fig. 2.4. It is thus much coarser than clay, giving a good packing arrangement in the body mixture.

Chapter 11 deals in more detail with the reactions that take place during firing, and the resultant constitution of the fixed body, but it may be helpful at this point to note certain features of the glassy phase. In the earthenware body, microscopic examination shows the presence of about 40-50 per cent quartz crystals, and about 10 per cent mullite crystals, embedded in a glassy matrix occupying about 45 per cent of the volume. From the known composition of the starting materials, it is possible to calculate the composition of this glass, and it is roughly as follows: $SiO_2$ 66 per cent, $Al_2O_3$ 28 per cent, alkalis 6 per cent by weight.

For a glass, this is rather high in alumina and low in alkalis, but it would melt at pottery firing temperatures, and its relatively high viscosity would be a hedge against excessive distortion. Bone china has a very different constitution. In that case, there would be in the fired body about 25 per cent of anorthite, and about 45 per cent of tricalcium phosphate, embedded in a glass of the following composition: $SiO_2$ 65 per cent, $Al_2O_3$ 15 per cent, alkalis 15 per cent, $P_2O_5$ 5 per cent.

In some bone china's the glass content may rise to nearly 40 per cent. It is, however, generally less than in the case of earthenware, which is somewhat surprising in view of the fact that the bone china is fired to a much lower porosity. The texture of the two systems is, however, very different. In some porcelains the glassy phase may occupy as much as 80 per cent of the volume.

In addition to the felspars, there are a number of other fluxes that can be used in pottery bodies. One of the more important is nepheline syenite, widely used in the glass industry and coming more into use in recent years for pottery bodies, especially for sanitary ware. It is found in Norway and North America. It is a complex material, consisting of a mixture of soda and potash felspars, together with the mineral nepheline. Nepheline, $Na_2O.Al_2O_3.2SiO_2$, is higher in alkali content than felspar, and very much higher than Cornish stone. Nepheline-syenite, therefore, is a rich source of alkali and has a vigorous fluxing action in bodies. There seems to be some difference of opinion as to whether it shortens or lengthens the firing range. A point to notice is that it makes a much lower contribution to the silica content of the body than do the other fluxes, so corrections to the recipe need to be made. Nepheline itself has a relatively high melting point; its fusion in bodies appears to be preceded by a reaction with available silica to form felspar.

When special characteristics are needed in bodies, fluxes can play an important part. For example, talc in small proportions can promote the formation of cordierite, giving bodies of low thermal expansion. In much higher proportions it can be used to give useful electrical properties, or to reduce moisture expansion. Low thermal expansion systems can be made by using lithia as a flux, derived from such minerals as lepidolite or petalite. A detailed description of the reactions developed in these bodies is beyond the scope of this book.

It may be useful to conclude this section by collecting together for easy reference the principal characteristics of the more commonly used fluxes that have been discussed above. This is done in Table 2.1 which shows the calculated composition for the pure minerals, and some typical analyses for commercially available raw materials.

**Table 2.1** – Chemical composition of fluxes

| | Minerals | | | Raw materials | | | |
|---|---|---|---|---|---|---|---|
| Oxide (%) | Felspars | | Nepheline | Potash felspar | Soda felspar | Cornish stone | Nepheline-syenite |
| | Orthoclase | Albite | | | | | |
| $K_2O$ | 16.92 | – | – | 11.9 | 2.2 | 4.6 | 9.8 |
| $Na_2O$ | – | 11.82 | 21.81 | 2.8 | 8.0 | 3.7 | 7.4 |
| $Al_2O_3$ | 18.31 | 19.44 | 35.89 | 19.0 | 19.9 | 12.0 | 24.0 |
| $SiO_2$ | 64.76 | 68.74 | 42.30 | 65.8 | 67.2 | 76.0 | 57.0 |
| $K_2O + Na_2O$ | 16.92 | 11.82 | 21.81 | 14.7 | 10.2 | 8.3 | 17.2 |

## 2.3 FILLER

A simple view of the filler in a pottery body is that its main role is to occupy the space left by the clay and the flux. That is an important function in itself, since the clay and flux together in some cases comprise only 50 per cent of the recipe. But that would be too simplistic a view.

The filler has a number of contributions to make in its own right. In so far as it may remain substantially unaltered in the firing, it provides a rigid skeleton giving stability and reducing contraction and distortion. In many cases it plays a significant part in the reactions taking place during firing and has a determining effect on the finished structure. Its own intrinsic physical properties often have their own contribution to make, and its relationship to the glassy matrix with which it is in contact influences many properties, such as whiteness, translucency, and strength. Add to this the fact that it needs to be available in large tonnages at a reasonable price, and it will be seen that it merits careful consideration. The principles that determine the selection of particular fillers for a given purpose have been outlined by Bruce and Wilkinson (1966), who also summarized the relevant properties of a wide range of possible materials.

### 2.3.1 Silica

Silica in one form or another is widely used as a filler in pottery bodies. One reason for this is its availability in large quantities at a commerical price; the British pottery industry uses up to 200 000 tons a year, and large quantities are used in other industries, notably glass. Since it comprises some 60 per cent of the Earth's crust, continuity of suppy would seem to be reasonably secure.

The most widely used form is that derived from sand or quartz rock. Most countries outside Britain have used this source for their traditional porcelains, but British practice until recently has been based on the use of flint. This material was first used in English earthenware early in the eighteenth century, primarily to give whiteness and to reduce plasticity and contraction. It can be obtained either in the form of siliceous nodules embedded in chalk, or as pebbles from nearby beaches. It needs to be calcined before grinding. It consists of very fine crystals of so-called cryptocrystalline quartz, about 1 per cent water, and very fine pores. When calcined, the water is driven off, and some change in crystal structure may occur. This weakens the bonding and makes the material easier to grind. In recent years, the use of sand has increased rapidly in Britain, and now probably represents more than half of the total silica used. Johnson (1976) has reported on some of the significant differences between flint and sand in pottery bodies.

Silica can exist in a wide variety of crystal structures. The study of its structure, and the influence of time, temperature, and impurities on the crystallization has received a great deal of attention and there is now an enormous literature on the subject. So far as whiteware bodies are concerned it is only necessary to consider two of the crystal forms, quartz and cristobalite, but an understanding of these two is crucial to a proper appreciation of their function in the body.

The characteristics of most significance are the thermal expansion, and the changes in crystal structure on heating, and especially the associated volume changes. Quartz exists at room temperature in the α-form with a specific gravity of 2.65. When heated to 573 °C there is an irreversible change to the β-form, with an increase in volume of about 1 per cent. Cristobalite, in the α-form, has a specific gravity of 2.32, and changes to the β-form at about 225 °C, with an increase in volume of about 3 per cent. The exact temperatures and the rates at which these inversions take place are influenced by the rate of heating, and the presence of other elements.

Sand and quartz rock consists mainly of well-crystallized quartz grains. Flint on the other hand, has a much smaller crystal size, and also contains some water, and some minor impurities. The specific gravity of raw flint is thus less than that of quartz, generally in the range 2.58 to 2.62. When heated to 1100 °C the value falls steadily to around 2.45; prolonged heating at higher temperature converts much of the quartz to cristobalite and the value drops to around 2.25. The quartz in sand can also be converted to cristobalite, but this takes place much less readily than in flint.

The presence of quartz or cristobalite in varying proportion has important repercussions in pottery technology. Among these may be noted

(a) The effect of density changes on body mixing.
(b) The importance of overall thermal expansion.
(c) The consequences of the crystalline inversions.

It is clear that where calcination brings about appreciable density changes, body proportions by weight will be altered where volume mixing is practised and vice versa. These variations are small but not insignificant.

The thermal expansion characteristics are, however, of vital importance in relation to the problem of glaze fit. The thermal expansion curves for quartz and

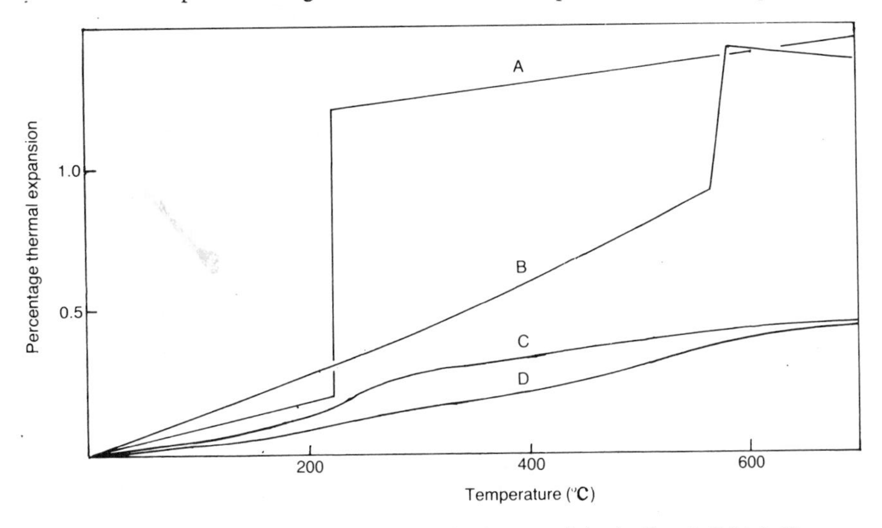


Fig. 2.5 – Thermal expansion of silica and silica-containing bodies. A, Cristobalite; B, quartz; C, body rich in cristobalite; D, body rich in quartz.

cristobalite are shown in Fig. 2.5, together with curves for two bodies, one of which is rich in quartz, the other rich in cristobalite. For most whiteware products, it is necessary to have the thermal expansion of the body higher than that of the glaze by about 0.05 per cent in order to counteract any possible body expansion owing to the adsorption of moisture. Most glazes have a notional softening point in the region of 500 °C, and it is the contraction of the body and glaze from this point downwards that determines the final glaze compression. Because the inversion of cristobalite takes place below 500 °C, whereas that of quartz is above 500° C, the presence of cristobalite helps to provide a higher expansion at the glaze softening point than is the case with quartz. This is clearly shown in the body expansion curves in Fig. 2.5. Most glazes have a thermal expansion of about 0.35 per cent to 500 °C, and it is not easy to produce glazes firing at normal temperatures with expansions much lower than this, say 0.25 per cent. It can thus be seen that any component that increases the body expansion is of great technical importance, and silica in the form of cristobalite is the best agent in this respect. Cristobalite may come, of course, in part from pre-calcination of sand or flint, or may be formed in the body during the firing process.

Over against the advantages associated with the use of crystalline silica as a filler, it is pertinent to note that there are several serious drawbacks, which have been discussed in detail by Dinsdale (1963). The first relates to the problem of dunting. This is the term generally used to denote cracking in the fired product, either in the kiln or subsequently, as a result of stresses associated with temperature gradients. The problem is particularly severe in the case of large complicated shapes such as are encountered in the sanitary ware or electrical porcelain industries. In any practical firing process, it is inevitable that temperature gradients will be set up between different parts of a piece of ware, both in heating and cooling. These temperature differences give rise to stresses and if these exceed the strength of the ware, cracking results. The higher the thermal expansion of the body, the greater will these stresses be, so the presence of silica is harmful in this respect. More than this, there is the possibility that two adjacent parts of the body may be at temperatures on either side of the crystal inversion point, and in this case the stresses are particularly severe. Of course, ceramic technologists have learned to live with this problem. Firing schedules are carefully adjusted so as to minimize the rates of heating and cooling, especially at the inversion temperatures, and so keep the dunting losses within reason. But bodies with lower expansion, and free from sudden volume changes, would offer significant rewards in terms of less loss and more possibility of rapid firing techniques.

A second result of this thermal behaviour of silica is that there is a mismatch between the body matrix and the silica particle. Examination of the body shows many large silica particles, unreacted, and in a state of strain *vis-a-vis* the surrounding body. Cracks frequently appear either in the particle or in the body, and these are responsible for a reduction in fired strength. Removal of the over-sized particles results in a substantial strength increase. Substitution by a filler with no inversions and closely matching the matrix would be an advantage in this respect.

Other disadvantages may be noted. It is now known that the presence of crystalline silica in the body has a harmful effect on glaze appearance. Many factors affect glaze appearance; strangely, the porosity of the body is not among the more important. The most important by far is the presence of bubble; these may remain in the glaze after firing, or the results of bubbles bursting may be apparent on the surface. Many of these bubbles are associated with particles of silica protruding from the surface of the body. Large silica particles have a much more serious effect than small ones. Bodies like bone china, relatively free from these particles, give much better glaze finish than bodies containing silica particles. Indeed, it is difficult to obtain good glaze finish at all in the presence of particles greater than about 50 $\mu$m in size.

The fact that large silica particles are associated with bubbles and cracks is also of key significance in relation to the spit-out fault that can occur in the decoration of porous bodies. This fault is caused by the desorption of moisture from the internal surfaces of the pores when the ware is heated up in the decorating kiln. A crack through or round a silica particle gives easy means of egress for the gas under pressure and a crater results.

It may finally be noted that silica is toxic when inhaled, and although risk to operatives can be obviated by suitable controls, silica-free bodies would offer an advantage in this respect.

All in all, then, crystalline silica has many disadvantages. The fact that these are outweighed by its more positive qualities ensures that its widespread use as a filler in whiteware bodies is likely to continue.

### 2.3.2 Alumina

The properties of alumina are such as to make it one of the most important ceramic materials. It has a high melting point, is resistant to chemical attack, and has desirable electrical and mechanical properties. For these reasons, it has long been used for refractory components, crucibles, spark plugs, and substrates in electronic devices. It can now be manufactured free from pores, and thus translucent, giving it a special function in lighting equipment.

In all these uses, alumina is used in its sintered form, with only minor additions of other materials. In relation to whiteware bodies, however, interest centres on its use as a filler as a total or partial replacement for silica. The attraction here is the inherent high strength and the absence of crystalline inversion. Alumina can occur in a number of crystal forms, but the $\alpha$-form is the one that is stable at high temperatures. Pure $\alpha$-alumina gives the best results in whiteware bodies, but for economic reasons consideration needs to be given by hydrated forms, which are widely available in bauxites and in association with other minerals. Both materials have been successfully used.

As far as strength is concerned, $\alpha$-alumina has a Young's modulus of $2.8 \times 10^{11}$ Nm$^{-2}$ compared with $6 \times 10^{10}$ Nm$^{-2}$ for flint. It is also free from the weakening effect of crystalline inversions associated with silica. It is thus possible to obtain whiteware bodies with double the strength, even when only 30 to 35 per cent is used in the mix. This high strength is important in some electrical porcelain products. It also confers a desirable degree of extra strength in some tableware products, especially where conditions of use present severe mechanical stress. There have been considerable developments in the use of alumina in tableware bodies over the last two or three decades.

There are, however, some disadvantages to be taken into account. The material is hard and therefore difficult to grind to the required degree of fineness. The cost is very much higher than that of any of the forms of free silica. The colour is not always as white as may be desired. The thermal expansion is appreciably lower than for flint or quartz, being not more than 0.4 per cent to 500 °C, thus making for some problems in finding suitable glazes. Body expansions greater than 0.3 per cent are not easy to achieve.

Another difficulty is that of increased weight. The specific gravity is 3.95 compared with around 2.5 for other materials. With 35 per cent present in the body, this represents an overall weight increase of 20 per cent. This is very undesirable in the case of tableware, both in the context of manufacture and of use.

The increased strength allows some reductions in thickness to be achieved, but there is an inevitable compromise between weight and strength.

In spite of all the disadvantages the use of alumina in whiteware bodies is likely to increase rather than diminish as further investigation into the possibilities proceeds: its use in whiteware bodies has been reported by Batchelor and Dinsdale (1960) and by Allen (1964). A detailed study of its effect on strength has been presented by Dinsdale and Wilkinson (1966).

### 2.3.3 Calcined clay

Another possible source of filler that would seem to merit consideration is calcined clay. China clay heated to over 1500 °C is converted to mullite, $3Al_2O_3.2SiO_2$, which is widely known as a refractory material under the trade name molochite. It has a specific gravity of about 3.03, and a thermal expansion coefficient of $4.5 \times 10^{-6}$, equivalent to 0.22 per cent to 500 °C. It is free from undesirable crystal inversions. It is likely to be expensive on account of the high calcination temperature, and the low thermal expansion would give bodies that were difficult to glaze.

The question arises, however, as to whether it might be possible to make a suitable filler from a less pure clay, with substantial free silica content, calcined at a much lower temperature, say around 1200 °C. The aim would be to convert some of the clay to mullite but also to form a glass which might contain minute crystals of cristobalite or quartz, thus keeping the expansion. With a careful selection of clays this can be done. Bodies with expansions appreciably higher than for those containing molochite can be formulated, but even so there are still likely to be some problems with glazing. A more formidable difficulty is that of finding suitable clays. Generally this means clays that are reasonable in cost because they are not suitable for other purposes. The supply problem is likely to inhibit the use of this type of filler.

### 2.3.4 Wollastonite

Calcium-bearing minerals offer some attractive features as fillers. Whiting is sometimes used in tile bodies to reduce firing contraction, and wollastonite offers potential as a constituent of low-loss electrical porcelain bodies. The use of wollastonite as a general filler for whiteware body systems has been examined in some depth in recent years.

Wollastonite, $CaO.SiO_2$, occurs in two principal crystal forms. The $\alpha$-form, often known as pseudo-wollastonite, is stable at temperatures above about 1125 °C. The $\beta$-form is stable below 1125 °C, and occurs naturally or may be readily synthesized by heating a mixture of finely ground sand and limestone to a reasonably low temperature. The $\beta$-form is the one of most interest to ceramists, although the $\alpha$-form does offer a higher thermal expansion. The $\beta$-form has an expansion of about 0.3 per cent to 500 °C which is just about tolerable from the glazing point of view. However, there are certain features of its behaviour in bodies that make for problems in manufacture. Bodies containing wollastonite tend to have a very short firing range, and there is often excessive reaction between the body and the glaze. Moreover, at biscuit firing temperatures, anorthite tends to result from interaction between wollastonite and clay, thus further reducing the thermal expansion. For these reasons, the use of wollastonite in whiteware bodies has not developed on a wide commercial scale.

### 2.3.5 Bone

Bone china, although made in some degree in other countries, has a distinctively British flavour. It has formed a major part of the British tableware industry for nearly two centuries. The particular characteristics that appeal are the whiteness, the translucency, the high glaze and decoration quality, and the high strength, and much of this derives from the use of bone as a constituent. As much as 40 or 50 per cent calcined bone may be used.

It is at first sight surprising to find this material in a ceramic body at all, since it is the only one that is not derived directly from the earth. Bone has been added in small quantities for many centuries, but the bone china system now in common use was established by Josiah Spode towards the end of the eighteenth century. No doubt its function was to give a white, translucent, appearance and the full implications of its reactions in the body would not be fully understood at that time.

The source of supply is mainly ox bone, and when this has been collected from the abbatoirs, most of the organic constituents are separately removed. There is a large and growing demand for the fats, glue, gelatin, and other constituents, and even the residual bone is becoming increasing used in fertilizers and foodstuffs. For use in the ceramic industry, the bone has to be calcined to around 1000° C, it being generally felt desirable for a small amount of organic matter to be left in, though too much of this can be troublesome.

The degree of calcination has a considerable influence on the crystal size, which in turn influences the size distribution on grinding. The calcined bone consists mainly of hydroxyapatite ($Ca_5.(PO_4)_3.OH$), though there are usually small but significant quantities of $Na_2O$, MgO, and carbonate. The Ca:P ratio is 1.66. The minor constituents are important in that they affect the fluid properties of slips containing bone. The chemistry is complex and beyond the scope of this book, but it is becoming increasing clear that such factors as ageing of slop bone, and the effect of atmospheric $CO_2$, can have a marked influence on the state of flocculation, which in turn is an important factor in the working properties of the body. In addition to its complex chemistry, it is also of interest to note that bone has a unique cellular physical structure, that is not matched by other sources of calcium phosphate that have been used as substitutes.

The function of bone in the body is quite different from that of other fillers. Whereas most of these remain partially or wholly unreacted during the firing, bone reacts with the other constituents, clay and stone or felspar, to produce a new assemblage of crystals in a glassy matrix in a manner that is described in more detail in Chapter 3. Suffice it to say at this point that the fired composition is roughly 25 per cent anorthite, 45 per cent tricalcium phosphate, and 30 per cent glass. The optical properties of these components is such as to give rise to the high degree of translucency of the body. The mechanical properties are also good, since there are no weakening crystalline inversions. With regard to the thermal expansion, the situation is rather complicated. The percentage expansions to 500° C are about 0.65 for tricalcium phosphate, and 0.25 for anorthite and the glass. In the event it turns out that the body has an expansion in the range 0.40 to 0.45 per cent, and there are thus no problems with regard to finding a suitable glaze.

Many attempts have been made to find substitutes for bone. Alternative sources might be rock phosphates or synthetic tricalcium phosphate, but in general the use of these materials results in problems either with the unfired body or during firing. Changes in colour, contraction, and vitrification range are all disadvantageous. Recent investigations have shown an interest in the use of a calcined mixture of dicalcium phosphate dihydrate and limestone (Taylor *et al.*, 1979). Other possibilities exist, but it seems likely that animal bone will continue to be the source of filler for bone china, at least until the pressure on supplies makes it necessary to use alternatives.

### 2.3.6 Principles of selection

The characteristic requirements of whiteware bodies have been summarized by Dinsdale (1976). There are no ideal fillers, and the selection of one for a particular body system inevitably means a compromise between a large number of characteristics, some of which are usually in conflict with each other. For products that have to be glazed, thermal expansion is of vital importance. Compatibility with clay, freedom from inversions, good colour, and low density all need consideration. When technical requirements have been met, practical economics demand assurance with regard to cost and continuity of supply.

## REFERENCES

Allen, C. J. (1964) *Tr. Brit. Cer. Soc.*, **63**, 77.

Bailey, R. T. (1974) *Tr. & J. Brit. Cer. Soc.*, **73**(8), 283.

Batchelor, R. W. (1973) *Tr. & J. Brit. Cer. Soc.*, **72**(1), 7.

Batchelor, R. W. and Dinsdale, A. (1960) *Trans. Int. Cer. Congress VII*, 31.

Bruce, R. H. and Wilkinson, W. T. (1966) *Tr. Brit. Cer. Soc.*, **65**, 233.

Dinsdale, A. (1963) *Tr. Brit. Cer. Soc.*, **62**, 321.

Dinsdale, A. and Wilkinson, W. T. (1966) *Proc. Br. Cer. Soc.*, No 8. 119.

Dinsdale, A. (1976) *Sci. of Ceramics*, VIII, 51.

Johnson, R. (1976) *Tr. & J. Brit. Cer. Soc.*, **75**(1), 1.

Mitchell, D. (1974) *Tr. & J. Brit. Cer. Soc.*, **73**(8), 287.

Royle, J. D. (1974) *Tr. & J. Brit. Cer. Soc.*, **73**(8), 291.

Taylor, D., Nijhawan, K. K. and Norris, A. W. (1979) *Tr. & J. Bri. Cer.*, **78**(5), 108.

# 3

# Design of body systems

In a historical sense the title of this chapter is misleading. Most of the body systems that constitute the mainstream products in the whiteware industry evolved over a long period of time, and their development arose out of a set of circumstances rather than a deliberate act of design. It had to be thus, since the potters of earlier days had to work more by instinct and intuition than by formal principles. They did not have available to them the vast mass of data and the increased scientific and technological knowledge that has emerged during the past few decades. The history of the development of particular body systems has been well reviewed in the case of bone china by Franklin and Forrester (1975); electrical porcelain by Johnson and Robinson (1975); and hard porcelain by Rado (1964, 1971, 1975).

The situation is very different today. There is a growing understanding of the properties of materials, of plastic behaviour, of reactions under heat, and of the constitution of fired materials, so that the reason for the various properties is better understood. In particular, the unravelling of the nature of the fired body, in terms of glassy matrix, crystal content, pores and flaws, has made it possible to relate fired properties to constitutional features in a quantitative manner. There is thus the beginning of the possibility of design in the absolute sense. Moreover, studies in high temperature reactions and the availability of phase diagrams covering a wide range of compositions, have made it possible to predict with a high degree of accuracy the changes that might follow from a change in either composition or temperature. One significant bonus from this is the elimination of a lot of tedious and expensive experimental work.

Nevertheless, body design is still difficult, not so much because of a lack of information but because the problem is intrinsically complex. We are confronted by the task of bringing together a large number of variables and constraints. The variables are not always independent of each other, and the constraints are frequently contradictory. The result is that no body system represents a complete solution of the problem; all fall short of the ideal in one or more respects. Since the ideal is not available, the problem is to optimize the situation so as to satisfy as many requirements as possible, as far as possible. Some constraints are much more crucial than others.

We can best approach the subject by dividing it into five sections, though these are not unconnected.

## 3.1 RAW MATERIALS

For any body system to be of practical interest, its components must be available in sufficient quantity at the right level of cost. It has already been noted that three types of material will usually be needed – clay, flux, and filler. All these materials must be free from intolerable impurities, and must give an acceptable colour when fired, except in the case where opaque glazes or engobes are to be used.

## 3.2 UNFIRED PROPERTIES

Many of the unfired properties of the body are determined by the clay content, and the relationship between the particle size of the clay and that of the non-clay materials fixes the packing density, which in turn affects the plasticity, strength, contraction, drying and contraction behaviour. For large-scale production, with minimum loss, the body properties must be such as to suit whichever forming technique is employed. Different characteristics are required for plastic making, casting, or dust pressing. The increasing use of automatic machinery and high rates of production means that the margin of error is reduced, and invariability of properties becomes almost as important as their absolute values.

After the article has been formed, fettling and handling processes also require appropriate levels of plasticity and strength. Drying shrinkage and distortion need to be kept within limits.

## 3.3 FIRING

Body systems must be chosen so that they mature at biscuit firing temperatures within reasonable practical limits, which will usually be between 1100 °C and 1250 °C for most whiteware products. For glazed ware, twice-fired, biscuit firing temperature needs to be at least 50 °C higher than that for glost. Excessive contraction or distortion during firing is not acceptable in products where dimensional accuracy is important.

## 3.4 FIRED PROPERTIES

Since the requirements of the unfired system usually determine the nature and quantity of clay and flux used, and since this automatically fixes the quantity of the filler, it is clear that the nature of the filler has a critical part to play in achieving the desired properties in the fired body. High modulus of elasticity, and fine grain size, will assist fired strength. Probably the most important requirement is high thermal expansion, since the range of options available in glazes at normal temperature is limited. If environmental considerations ever restrict the use of lead in glazes, this aspect will becomes even more difficult than it is now. The contribution of the filler to the desired fired properties is sometimes achieved by reaction with the flux and clay, while in other systems the filler may retain most of its original identity. An important point to note is that in many body systems more changes can be brought about by changes in temperature or time of firing than by changes in composition. On the other hand, many potentially interesting body compositions are ruled out because of the difficulty in glazing.

## 3.5 THE PRODUCT IN USE

In the last resort, what determines the success or otherwise of a whiteware product in the market-place is fitness for service. The demands made on the product in different sectors are very different. Many of them are difficult to quantify, but they are no less real for that, and must be taken into account in body design. Dinsdale (1968) has suggested a rough guide to some of the more relevant ones, as shown in Table 3.1, where an attempt is made to assess their relative importance, high numbers registering a high degree of significance.

**Table 3.1** – Importance of various properties of whiteware products

| | Tableware | Cooking ware | Wall tiles | Sanitary ware | Electrical porcelain |
|---|---|---|---|---|---|
| (i) Accuracy of dimension | 3 | 2 | 5 | 3 | 5 |
| (ii) Porosity | 3 | 2 | 0 | 4 | 5 |
| (iii) Strength | 3 | 3 | 1 | 3 | 4 |
| (iv) Acoustics | 2 | 2 | 0 | 0 | 0 |
| (v) Thermal shock | 0 | 5 | 0 | 0 | 2 |
| (vi) Electrical | 0 | 1 | 0 | 0 | 5 |
| (vii) Crazing resistance | 5 | 5 | 5 | 5 | 5 |
| (viii) Appearance | 5 | 3 | 3 | 3 | 3 |
| (ix) Glaze surface | 5 | 3 | 3 | 3 | 3 |

We consider each of these properties in turn.

(i) Articles made by traditional pottery processes, using traditional materials, always change size and generally change shape during manufacture. The variation in size from article to article depends on the product concerned, and may or may not be of interest to the consumer. Wall tiles achieve a high degree of dimensional accuracy, and the emergence of the constant size tile is a great technological advance, since it eliminates sorting in the factory and at the same time improves the ease of fixing on the wall. Many standards for wall tiles require the linear dimensions within a batch to vary by not more than ±¼ per cent. It is of interest to compare this with tableware, where the variation in the size of plates is usually about ±1¼ per cent, though some factories work to about half this and others to half as much again. Published requirements for sanitary ware are more of the order of ±2 per cent. It is interesting to speculate how far dimensional accuracy derives from consumer demand and how far from technological possibility; probably a bit of both. On the latter side, contraction, which mainly determines finished size, can be related to firing temperature, and the degree of control required can easily be calculated from the slope of the temperature-porosity curve. It is of significance that the size tolerances associated with wall tiles, earthenware, and vitreous ware correspond to the fact that these products are fired to the top, middle, and bottom of the vitrification curve respectively.

Dimensional accuracy involves consideration of the packing of the raw materials, drying shrinkage and vitrification rate, and is thus a complex phenomenon. In addition to the linear dimension, there is the question of distortion, mainly associated with glass formation in the body. The compromise between minimum distortion and optimum vitrification is always difficult to achieve, and the problem is as complex as that of reducing variation in linear dimension.

(ii) Apart from the effect it has on other properties, porosity is significant in itself, especially in the case of products where hygiene is important. The effects are, of course, minimized by glazing. In any case, the complete elimination of a small amount of residual porosity is almost impossible, except under very special manufacturing procedures.

(iii) A considerable amount of effort in recent years has gone into the study of the strength of ceramics, and an appreciable advance in understanding has resulted. But it is easy to over-estimate its significance in relation to whiteware, and the basic requirements are very different for the different products. Some electrical porcelain components have to operate under conditions of severe mechanical stress, both static and dynamic, and adequate strength is essential for the fulfilment of function. No great problem exists in the case of sanitary ware, and wall tiles are only rarely required to withstand impact. Impact strength is the problem with tableware. Service treatment is often so rough that marginal increases in strength are not enough to prevent fracture under catastrophic conditions. In improving the resistance to breakage of tableware, attention has to be paid to all aspects of the materials used and the texture of the fired body; correct glazing can also play a part.

(iv Acoustic properties are not usually regarded as important, but it is worth mentioning that handling large quantities of crockery can be a noisy operation. Unfortunately, the low damping factor is an inherent characteristic of the material, and the situation is worsened by harder firing.

(v) Too much emphasis has often been laid on thermal shock. Tableware and sanitary ware will easily withstand contact with boiling water, and neither should be expected to be subjected to more severe treatment than this, though both will, in fact, stand more. Some electrical porcelain products are subject to severe thermal shock, sparking plugs being a well-known example. In the field of cooking ware the property is of great importance, especially if the utensil is to be in contact with a gas ring or hotplate. Thermal shock resistance is, of course, a secondary property, depending mainly on thermal expansion, strength, elastic modulus, and porosity.

(vi) Electrical resistivity, dielectric constant, loss, and breakdown strength are of high significance in many electrical porcelain applications, but of no consequence in other whiteware fields. The requirements are often stringent, and a considerable degree of control is needed. Porosity can again be an important feature.

(vii) Crazing is not now a serious problem. The increasing understanding of moisture expansion and the improved control of glaze body fit have virtually eliminated the fault. Nevertheless, attention needs to be paid to all the factors involved in order to keep the risk at a minimum.

(viii) For many whiteware products, customer appeal is more related to appearance than technical quality. Good overall colour is important, especially in the near-white region, and irregularities such as specking are not to be tolerated. The purity of the raw materials is paramount.

(ix) Nearly all whiteware products are glazed, and the quality of the final glaze surface determines gloss and visual appeal. One of the more obstinate problems is to reduce scratch and abrasion damage, since these properties do not vary much over a wide range of glaze compositions.

## 3.6 FEATURES OF SOME BODY SYSTEMS

All the above considerations emphasize the fact that body formulation is a very complex exercise. Attempts have been made in recent years to involve the use of computers. The idea is to feed in all the known variables and the requirements, and to find an optimum solution. This may simplify the problem in limited areas of application, but it does not resolve the underlying dilemma in that many of the elements in the pattern of possibility and requirement are contradictory. The ceramist has to live with compromise, but there are aids, not least among them the use of phase diagrams. These are now available for a very wide range of ceramic compositions. They show what liquid or solid phases are present in equilibrium for particular compositions and temperatures. They are of great importance in high temperature technology, such as refractories and special ceramics, where equilibrium is often obtained. Whiteware systems, at normal firing temperatures, however, are not in equilibrium. This technology is sometimes described in terms of arrested or incomplete reactions. This does not mean that phase diagrams cannot be used; on the contrary, they can often furhish a very useful guide to the nature and quantity of crystals and liquid in a given situation, and what is sometimes more important, to the probable effect of changes in composition or heat treatment.

Phase diagrams can be constructed in a variety of ways, but the most useful for our purpose is the triangular diagram, and especially a slice taken at a particular temperature. It is beyond the scope of this book to deal with this subject in detail, but an indication of the approach is not out of place. As has already been indicated, most of the whiteware systems under consideration can be thought of as consisting of three components, namely, clay, flux, and filler. These are known as triaxial compositions. Typical recipes for some well known products are shown in Table 3.2.

**Table 3.2** – Typical triaxial compositions

| | Earthenware or semi-porcelain | Bone china | Hard porcelain | Sanitary whiteware | Sanitary fireclay | Wall tile | Electrical porcelain |
|---|---|---|---|---|---|---|---|
| **Clay** | | | | | | | |
| Ball clay | 25 | | 10 | 23 | | 22 | 25 |
| China clay | 25 | 25 | 40 | 24 | | 22 | 25 |
| Fireclay | | | | | 60 | | |
| **Flux** | | | | | | | |
| Stone | 15 | 30 | | | | 15 | |
| Felspar | | | 20 | 20 | | | 25 |
| **Filler** | | | | | | | |
| Quartz/flint | 35 | | 30 | 33 | | 41 | 25 |
| Bone | | 45 | | | | | |
| Grog | | | | | 40 | | |

### 3.6.1 Clay : felspar : quartz

The earthenware body is in the triaxial system clay : stone : flint. The general features of this system are shown in Fig. 3.1(a), which gives an indication of the effect of changes in composition in the three directions. Reservations have to be kept in mind since stone is not wholly felspar, nor is clay wholly kaolinite, but this does not invalidate the general argument. To predict what is likely to happen when a body in this system is fired, we consider the corresponding phase diagram. This would be that for felspar ($K_2O.Al_2O_3.6SiO_2$), metakaolin ($Al_2O_3.2SiO_2$), and quartz ($SiO_2$). This is part of the diagram for leucite ($K_2O.Al_2O_3.4SiO_2$), mullite ($3Al_2O_3.2SiO_2$) and quartz, and, more basically, within the system $K_2O$:$Al_2O_3$:$SiO_2$. All these diagrams are readily available. A study of the felspar:clay:quartz diagram shows that at whiteware biscuit firing temperatures the fired body consists of very little residual felspar, some mullite and a good deal of quartz.

The firing process can be described in these terms. The diagram shows that felspar melts at 1140 °C, but much happens before that temperature is reached. Pore water and adsorbed water are driven off by 250° C, and by 500° C the organic matter has been burned out and the clay molecule has given up its combined water. The quartz inversion follows at 573° C, and at 1000° C shrinkage begins, the body still having a high porosity. From then on to the peak firing temperature, felspar begins to react with other components and glass begins to form. The surface tension of the glass draws the particles together and reduces the porosity. This process is slow because the glass has a high viscosity, and becomes even more viscous due to the solution of silica. The fact that this vitrification process is slow is something of an advantage in practice, as it makes for a reasonably long firing range. As the firing proceeds, more and more glass is formed, and mullite crystals begin to develop, giving increased strength. Some solution of quartz grains takes place. How far this process is allowed to proceed determines the constitution of the fired body. At the earthenware end of the scale there is considerable residual open porosity, the quartz is largely unaltered, there is some felspar left, and not much mullite appears. At higher temperature, felspar disappears, open porosity is reduced almost to zero, though a small percentage of closed pores always remain, more glass is formed and more mullite crystals appear. At the much higher temperatures at which hard porcelain is fired, around 1400 °C, there is much more mullite and much more solution of the quartz. This lowers the thermal expansion of the body and presents the problem of low expansion glazes. This is overcome by the practice of biscuit firing the body to a relatively low temperature, 900–1000 °C, and then firing the glaze and body together to the peak temperature.

The fact that the degree of vitrification can be controlled by temperature and time does not mean that composition changes can be ignored. The area of practical triaxial compositions is shown in Fig. 3.1(b). Within that area wall tile

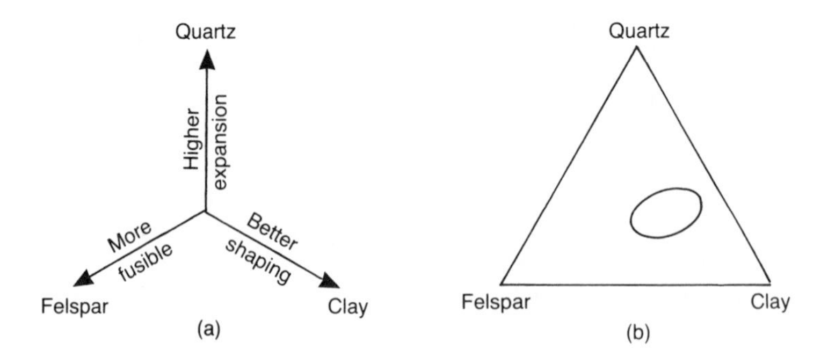


Fig. 3.1 – (a) General features of triaxial compositions. (b) Whiteware body composition.

and earthenware compositions are located towards the top right-hand side; vitreous sanitary towards the centre; and hard porcelain towards the bottom, a sequence of increasingly fusible compositions because of the increased felspar content.

### 3.6.2 Alumina bodies

A similar area is shown in Fig. 3.2(a) for bodies containing alumina (Batchelor and Dinsdale, 1960). The flux consists of 3 parts felspar and 1 part talc, and the bodies vitrify to zero apparent porosity at 1200 °C, over the range of compositions indicated. In this case, the glass is formed by reaction between the flux and the clay, the alumina remaining virtually unchanged.

### 3.6.3 Bone china

Bone china is a very different system, having unique constitutional features. The usable compositions are confined to the area shown in Fig. 3.2(b), although compositions outside this area are occasionally quoted in the literature. The relatively low clay content means that the plastic properties are less favourable than with earthenware, so the potting problems are likely to be more difficult.

What happens in the firing is somewhat complex, but is now more clearly understood than in earlier days (Dinsdale, 1967). It was known half a century ago that bone added to a body in small quantities acted as a flux, having a maximum fusion effect at 15 to 17 per cent. At higher proportions it tended to become more refractory. In the much larger quantities used in commercial bone china bodies it was not clear what reactions were taking place. A combination of modern investigative techniques cleared up the mystery. It was shown by x-ray studies on various combinations of the three components that the fired body consisted of two crystalline phases, namely anorthite ($CaO.Al_2O_3.2SiO_2$) and $\beta$-tricalcium phophate ($3CaO.P_2O_5$), embedded in a substantial amount of glass. Studies of the distribution of elements in the body confirmed this finding,

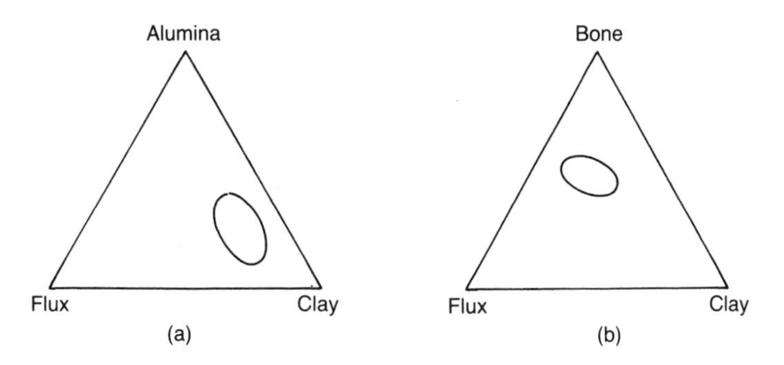


Fig. 3.2 – (a) Vitrification area for alumina bodies at 1200 °C. (b) Composition of bone china.

and phase diagram studies on bodies fired to 1500°C also gave the same conclusion. Moreover, the x-ray studies at temperatures early in the firing indicated the nature of the reactions that were taking place. Briefly, these are as follows. All the metakaolin from the clay reacts with some of the calcium from the bone to form anorthite. The remaining calcium and the associated phosphate remain as β-tricalcium phosphate; the residual phosphate combines with the melted felspar to form the glassy matrix. The felspar plays no part in the reaction between the clay and the bone, but plays a crucial role in the reduction of porosity and the development of translucency. In the light of the information thus revealed it is possible to work out quantitatively the phases present in the fired body. It is worthwhile doing this in some detail, since the system is so different from the clay–felspar–quartz body, and since it well illustrates the method.

All the data required for this calculation are given in Table 3.3. We begin with the simplifying assumption that the clay may be represented by kaolinite and the bone by hydroxy-apatite, these containing approximately 14 and 2 per cent combined water respectively. It we take the nominal recipe clay 25%; bone 50%; felspar 25%, and remove the combined water, this reduces to metakaolin 22.5%: apatite 51.3%: felspar 26.2%, for the composition of the fired material. Using the known molecular weights, we can calculate the proportions by weight of the relevant oxides, as shown in Table 3.3.

**Table 3.3** – Constituents of bone china body

| | | $CaO$ | $P_2O_5$ | $Al_2O_3.2SiO_2$ | $Al_2O_3$ | $SiO_2$ | $K_2O$ |
|---|---|---|---|---|---|---|---|
| Apatite | M.W. | 560.8 | 425.82 | – | – | – | – |
| $10CaO.3P_2O_5$ M.W. = 986.62 | % | 56.84 | 43.16 | – | – | – | – |
| 51.3 parts | Parts | 29.16 | 22.14 | – | – | – | – |
| Metakaolin | M.W. | – | – | 222.14 | 101.96 | 120.18 | – |
| $Al_2O_3.2SiO_2$ M.W. = 222.14 | % | – | – | 100 | 45.90 | 54.10 | – |
| 22.5 parts | Parts | – | – | 22.5 | 10.33 | 12.17 | – |
| Felspar | M.W. | – | – | – | 101.96 | 360.54 | 94.20 |
| $K_2O.Al_2O_3.6SiO_2$ M.W. = 556.70 | % | – | – | – | 18.32 | 64.76 | 16.92 |
| 26.2 parts | Parts | – | – | – | 4.80 | 16.97 | 4.43 |
| Anorthite | M.W. | 56.08 | – | 222.14 | 101.96 | 120.18 | – |
| $CaO.Al_2O_3.2SiO_2$ M.W. = 278.22 | % | 20.16 | – | 79.84 | 36.65 | 43.19 | – |
| β-tricalcium phosphate | M.W. | 168.24 | 141.94 | – | – | – | – |
| $3CaO.P_2O_5$ M.W. = 310.18 | % | 54.24 | 45.76 | – | – | – | – |

Molecular weights: $CaO$ = 56.08; $P_2O_5$ = 141.94; $K_2O$ = 94.20; $Al_2O_3$ = 101.96; $SiO_2$ = 60.09.

Since all the metakaolin reacts with some of the calcium from the apatite, we have 22.5 parts of $Al_2O_3.2SiO_2$ combined with $(20.16 \times 22.5)/79.84 = 5.68$

parts of CaO to form 28.18 parts of anorthite. This leaves 29.16 − 5.68 = 23.48 parts of CaO to form tricalcium phosphate. So 23.48 parts of CaO plus (45.76 × 23.48)/54.24 = 19.81 parts of $P_2O_5$ make 43.29 parts of tricalcium phosphate. This leaves 22.14 − 19.81 = 2.33 parts of $P_2O_5$ to combine with the felspar to form the glass. The composition of the glass is thus

| | Parts | Percent |
|---|---|---|
| $SiO_2$ | 16.97 | 59.5 |
| $Al_2O_3$ | 4.80 | 16.8 |
| $K_2O$ | 4.43 | 15.5 |
| $P_2O_5$ | 2.33 | 8.2 |

and the constitution of the fired body is approximately; anorthite 28%: β-tricalcium phosphate 43%: glass 29%. Microscope and x-ray studies confirm that these proportions are roughly correct. They also show that the body is inhomogeneous, and, in particular, that the composition of the glass varies widely from point to point. It is of interest to note in this respect that if the three components are preconstituted, made up into a body and fired, the result is a much more homogeneous fired state, with appreciably superior physical properties. The process would, of course, be hopelessly uneconomic. The relationship between the glass and the crystal phases, and the gradual elimination of open pores during the firing, have an important bearing on such properties as strength and translucency. This subject will be dealt with in more detail in a later chapter. Suffice it to say at this point that it is one of the frustrating aspects of ceramic technology, that the optimum properties hardly every develop at the same time and temperature, so the best compromise has to be accepted.

### 3.6.4 Other bodies

The clay-felspar-quartz and the clay-felspar-bone systems represent the bulk of the tonnage of commercially produced pottery, but there are many interesting variations. Stoneware has a long history, and is traditionally made from a single clay, relatively refractory but containing enough flux to vitrify it at temperatures of 1200 to 1250°C. Coloured bodies have often been made from high iron content clays, especially in the teapot field. These coloured bodies are also effectively used for self-decorative purposes in the artware field. Special body systems that have gained world-wide renown are jasper, which is vitrified, and has a smooth surface, even when unglazed, and which may be coloured either green or blue; and basalt, so-called after the black vitreous volcanic rock.

## 3.7 MIXING

We end this chapter with a cautionary word about mixing. Once a body has been chosen for a particular purpose, and constituents selected to give the right fired properties, there remains the problem of how to bring them effectively together into a homogeneous mix. Two principal methods are used; wet and so-called dry mixing. Wet mixing has many advantages. It makes for a much more thorough distribution of the ingredients, and it makes sieving and magnetting easier. The subsequent filter pressing also has the effect of removing potentially harmful soluble salts. A disadvantage is that if the body is to be worked in the plastic state, or even for pressing, the required removal of surplus water is wasteful of energy. Dry mixing appears to be replacing wet mixing to an increasing degree, and has many advantages. However, for both methods the possibility of error in the proportioning of the materials is very real. A study of the true composition of bodies compared with the nominal recipes would be very revealing.

In wet mixing, the ingredients are made up into separate slips, the pint weights of which are measured. The solid content is calculated using an assumed value for the density. Since the density may vary from the nominal value, an appreciable error can creep in, so that when the slips are mixed together the proportions of the solids may not be as required. Apart from the assumptions relating to the density, the measurement of the slip density on such a small sample as a pint may not be sufficiently accurate. This disadvantage is often overcome by measuring both the volume and the weight of the slip in bulk.

With dry materials, it needs to be remembered that very few materials are at zero moisture content. Allowance still needs to be made for the moisture, and there are difficulties in representative sampling.

So care needs to be taken to ensure that the mix actually used is the one intended. Fortunately, the errors are minimized where a standard recipe is being used for long production runs, using the same raw materials.

## REFERENCES

Batchelor, R. W. and Dinsdale, A. (1960) *Tr. Int. Cer. Congress VII,* 31.
Dinsdale, A. (1967) *Sci. of Ceramics,* **III**, 323.
Dinsdale, A. (1968) *J. Brit. Cer. Soc.,* **5**(2), 96.
Franklin, C. E. L. and Forrester, A. J. (1975) *Tr. & J. Brit. Cer. Soc.,* **74**(4), 141.
Johnson, P. and Robinson, W. G. (1975) *Tr. & J. Brit. Cer. Soc.,* **74**(4), 147.
Rado, P. (1964) *J. Brit. Cer. Soc.,* **1**(3), 417.
Rado, P. (1971) *Tr. Brit. Cer. Soc.,* **70**(4), 131.
Rado, P. (1975) *Tr. & J. Brit. Cer. Soc.,* **74**(5), 153.

# 4

# Physical properties of mixtures

The physical properties of pottery body systems are interrelated in a very complex manner, but it is possible to establish certain basic principles by considering some of the elementary characteristics of mixtures of particles of different sizes. The way in which these particles pack together can be shown to have a profound effect on a number of very important technological factors. It is not always possible to derive exact quantitative relationships, but it is of particular help if the results of changes in body composition can be predicted in a general sense. In this way, much tedious experimental investigation can be avoided.

Body systems are not usually chosen so as to give a maximum density of packing, so that it is to be expected that there will be a wide range of unfired and fired properties among commercially produced pottery. Nevertheless, Dinsdale and Wilkinson (1966) have shown that certain general relationships can be established which are applicable to all systems.

In order to develop the concept of packing in this context it is necessary to make a number of simplifying assumptions. Some of these might at first seem naive, and might appear to make the model look less and less like the real system, but they can perhaps be justified by the nature of the conclusions arrived at.

The first assumption is that the particles are spherical, since this simplifies the theoretical analysis of the packing arrangements. The assumption ignores the fact that clay particles are known to be flat and plate-like, though they do often aggregate into clusters which are more equidimensional. Ground materials, such as flint, stone, and bone are reasonably spherical. Of all the possible geometrical shapes, spheres do not give the best packing arrangement; it is important to note that the inevitable presence of interparticle spaces, or pores, is of crucial significance in pottery technology.

The second assumption is that any given material can be represented by a single size parameter. In fact, both naturally occurring materials like clays, and ground materials, have a size distribution covering a fairly wide range of sizes. There are a number of ways in which a size distribution curve can be represented by a single number; a convenient one, used here, is a medium size above and below which there are equal weights of the material.

It is assumed, thirdly, that, in the industrial processes under consideration, the materials pack in a manner closely related to the ideal arrangement. Clearly, optimum packing is never quite achieved, but the model may still be useful in establishing limiting conditions.

Fourthly, it should be noted that most body systems contain three or more constituents. For the purpose of this analysis two-component systems are considered, it being assumed that materials can be arranged in groups so that they fall into two main categories. For example, in earthenware, the fine fraction would include ball clay and china clay, and the coarse fraction would include flint and stone. It is, of course, possible to extend the theoretical treatment to three or four sizes, but this would not result in any material improvement in understanding the technology of clay-based pottery systems.

It should also be noted that in considering packing arrangements the volume proportions of the system are important. Where materials of similar density are concerned, the weight proportions are similar, but when materials with very different densities, such as alumina, are introduced, this will not apply.

With these assumptions in mind, it is appropriate to consider the theory of the packing of a system of spheres of two different sizes. The analysis seeks to answer the following questions. How does the porosity vary with the size and proportions of these two fractions? Does this help to explain the unfired porosity found in pactice in different kinds of pottery bodies? If so, can the analysis be usefully extended to other important properties which might be expected to be influenced by unfired porosity, such as contraction, unfired strength, plasticity, casting rate and permeability, and even fired strength?

## 4.1 PACKING OF TWO-COMPONENT SYSTEMS

In considering two-component systems, the packing of spheres of the same size should first be noted. There are many possible regular formations, the closest being the rhombohedral, with a porosity of 25.95 per cent. At the other extreme, the cubic arrangement is the most open, with a porosity of 47.64 per cent.

Consider now two components of different size, mixed in different proportions. An important parameter is the ratio of the two sizes; the packing behaviour depends on this ratio rather than on the absolute size values. Each component has a porosity of its own. In practice this is found to be in the range 40 to 50 per cent, regardless of the absolute size. In the present case we assume it to be 40 per cent. In Fig. 4.1 the porosities of various mixtures are shown for a range of size ratios. The general characteristics of these curves can be deduced from theoretical considerations, but they are also well supported by experimental data on a range of real materials (Furnas, 1928). On the right-hand side of the diagram is represented 100 per cent of the coarse material with a porosity of 40 per cent. As the smaller component is progressively added it fills up the interstices and thus reduces the porosity. However, the packing eventually deteriorates and the porosity increases until it approaches that of the fine component. The best packing occurs at more than 50 per cent of the coarse component, and moves to higher percentages of coarse as the size difference becomes greater. It can also be seen that better packing can be achieved when the sizes are widely different than when they are nearly the same. Both these observations are of great practical significance when considering pottery body systems.

The practical use to which such a diagram can be put, may be illustrated by considering two tableware bodies of markedly different composition, namely, earthenware and bone china. A typical earthenware body might contain about 50 per cent clay, roughly divided between ball clay and china clay. The ball clay would have a somewhat finer grain size than the china clay, but taking them together a reasonable figure for the median grain size would be 1 $\mu$m. The other 50 per cent of the body would consist of a mixture of stone and flint with a median grain size of about 15 $\mu$m. Thus for this body, the parameters would be

Proportion of coarse component = 50 per cent
Size ratio fine : = 1:15 = 0.07.

This system is thus represented on Fig. 4.1 by the point A.

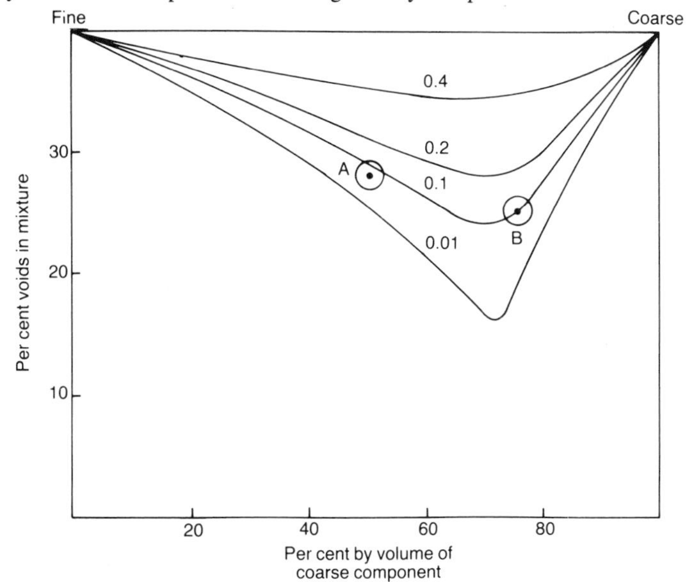


Fig. 4.1 – Porosity of two-component mixtures.

Bone china, on the other hand, would have only about 25 per cent clay content, again with a median size of about 1 μm. The rest of the body would consist of about 50 per cent bone and 25 per cent flux, the bone being finer than the flux, with the mixture having a mean size of about 10 μm. In this case,

Proportion of coarse component = 75 per cent
Size ratio fine : coarse = 1:10 = 0.1.

This body is represented on Fig. 4.1 by the point **B**.

It will be seen that points **A** and **B** are on different sides of the minima on the graph, and two important conclusions emerge. In the first place, both bodies behave similarly in that an increase in the difference between the sizes, that is a lowering of the size ratio, results in a drop in the porosity resulting from an improvement in the packing. But, secondly, increasing the proportion of the coarse material improves the packing in the case of earthenware, but makes it worse in the case of bone china.

It can thus be seen that in order to be able to predict likely changes in unfired porosity owing to changes in the proportion or grain size of the constituent materials, it is first necessary to establish the position representing the body system on a chart such as Fig. 4.1.

Another important practical point to note is that fine grinding of the body,

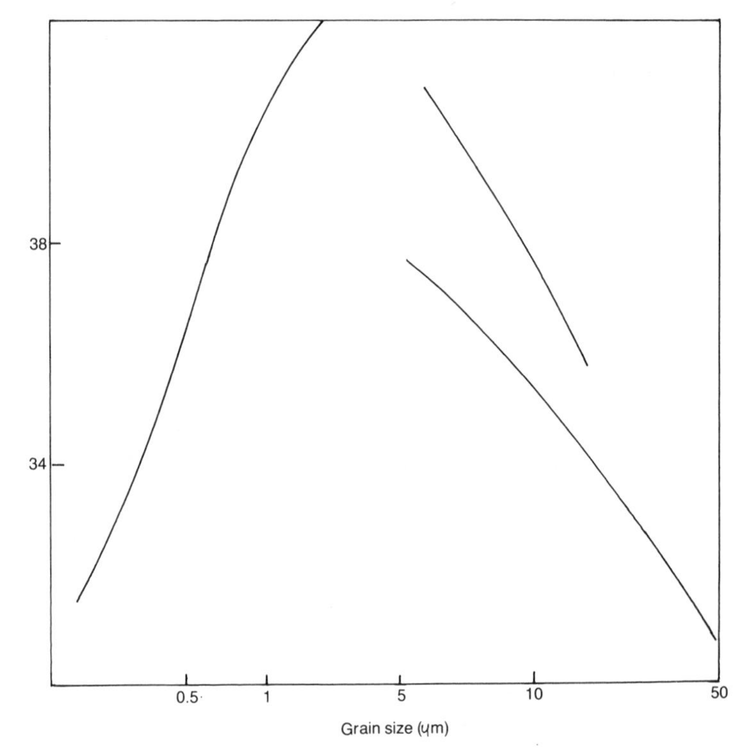


Fig. 4.2 – Effect of grain size on unfired porosity of body mixtures.

which may reduce both sizes equally, does not necessarily bring about better packing. Indeed, the probability is that it will cause more size reduction in the coarse material, thus increasing the fine : coarse ratio and worsening the packing.

## 4.2 EXPERIMENTAL VERIFICATION

The general trends indicated by the foregoing analysis can be experimentally tested in real ceramic systems in a number of different ways, some of which are illustrated in Fig. 4.2. Graph **A** represents the results of a series of experiments on three-component systems, with the clay content fixed at 50 per cent, and the size of the non-clay fraction varied between 6 and 18 $\mu$m. As the size of the non-clay material is increased, with a decreasing fine : coarse ratio, the packing is improved and the porosity is reduced, as would be expected from the theory. The results for another system, in which the fine end of the size range was fixed and the coarse end varied, are represented by Graph **B**. Again the proportions are kept constant, but the size of the 35 per cent alumina fraction is varied between 5 and 50 $\mu$m. The expected fall in porosity with increasing alumina size is confirmed.

An alternative method of changing the size ratio is by fixing the coarse end and varying the size of the fine fraction. This is illustrated in Graph C which indicates results from a body system in which the coarse materials were kept constant, but the 50 per cent clay content was made up from clays with a medium grain size ranging from 0.2 to 2 $\mu$m. As the size of the clay increases, the porosity increases. These practical results confirm the prediction from theory. In these body systems, packing can be improved by either increasing the filler size or decreasing the clay size, either of which increases the disparity between the sizes.

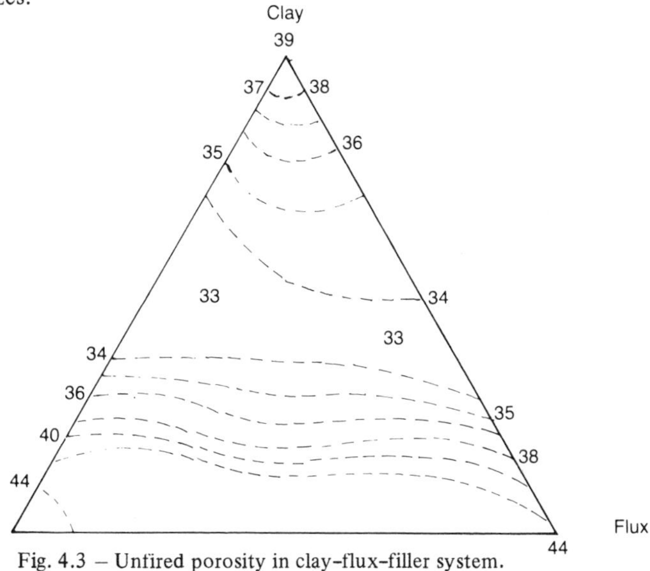


Fig. 4.3 – Unfired porosity in clay–flux–filler system.

It should be noted that fine materials sometimes aggregate into groups which themselves have a porosity. If these aggregates are not broken up in processing they will cause the porosity of the mix to increase. If the internal porosity of the aggregate is $P_2$ and the aggregate size causes it to pack with the other materials to a porosity $P_1$, then the resulting porosity of the mixture will be $P_1 + P_2(100 - P_1)/100$. In the examples considered so far, the proportions of the components have been kept constant and the sizes have varied. The other way of changing the packing arragement is to keep the size constant and vary the proportions. Fig. 4.3 shows the measured unfired porosities in a series of mixtures of clay : flux : filler. The model would predict that along the line joining two components the porosity would pass through a minimum and that there would be an optimum packing area somewhere in the centre of the triangle. This is confirmed by the results shown.

It is thus seen that a theoretical analysis of packing can be a valuable guide to the effect of various kinds of body changes on the unfired porosity. This conclusion must be qualified somewhat by remembering that the porosity can be modified to some degree by the forming process, for example by the moisture content in extrusion or by the pressure in dust pressing.

The theory can, of course, be extended to a consideration of the packing of three-component systems, but little practical advantage is to be gained by applying this to traditional pottery body systems. It is of more interest to enquire what other practical use can be made of the relationships so far established. Since the unfired porosity can be predicted from a knowledge of the characteristics and proportions of the constituent materials, it should be possible to derive valuable information about other important technological properties that might be expected to be related to unfired porosity.

## 4.3 WET-TO-DRY CONTRACTION

During the manufacturing processes a pottery article undergoes an appreciable reduction in size. Where dimensional accuracy is important in the finished article, a reliable estimate of the contraction is necessary in order to ensure that the mould or die is the right size. Even where this does not apply, other considerations, such as the fit of patterns in decorated tableware, dictate that control of size should be established. The overall contraction can be conveniently divided into two parts. When the article is fabricated in the mould, it generally has a finite moisture content, and when this moisture is removed in the drying process, a contraction may take place, known as the wet-to-dry contraction. Later in the processing another contraction occurs when the article is fired, and this is known as the dry-to-fired contraction. Considering at this point only the wet-to-dry contraction, it might be thought that there ought to be a direct relationship between the contraction and the unfired porosity, but it turns out to be not as simple as that. If a wide range of body compositions is examined and the moisture contents at fabrication are plotted against the wet-to-dry contractions, no clear correlation between the two is found. The reason becomes apparent if we consider the moisture content to consist of two parts. The first part is that which is required to fill all the pores. Generally, however, most fabrication processes need more water than this, so that there is a second part of the moisture content, which may be thought of as lubricating water. The amount of lubricating water required is roughly constant for bodies of quite different pore structures and hence quite different total moisture contents. Only this second part contributes to the contraction, since the first part can be removed by drying without any change in size, as all the particles are in contact.

If a wide range of bodies, with varying porosities, is studied it is found that there is a good correlation between the moisture content required to give the necessary level of plasticity for plastic working processes, and the unfired porosity. As the unfired porosity increases, so does the moisture content at making, in a fairly linear manner. It turns out that the line representing this relationship is a line of constant contraction.

Some theoretical considerations indicate why this should be so, and also enable a chart to be constructed for use as a practical guide to the prediction of contraction in a given body system. When dealing with either moisture contents or contractions it is necessary to define clearly the basis on which numerical values are calculated. It is proposed here to express moisture content as a percentage of the weight of the wet material (wet basis), and contraction as a percentage of the size of the wet article.

The various parameters involved can be related to each other in the following way:

Let $V$ = wet-to-dry volume contraction, as a percentage of the wet size,
$M$ = moisture content, wet basis,
$\rho$ = specific gravity of solid material,
$p$ = unfired porosity, per cent, i.e. the volume of the pores as a percentage of the total volume of the dry piece.

Then it can be shown that

$$M/100 = [100V + p(100 - V)] \div [100V + p(100 - V) + \rho(100 - p)(100 - V)]$$

and

$$V/100 = [M\rho(100 - p) - p(100 - M)] \div (100 - p)\ [100 + M(\rho - 1)]$$

In many practical cases it is more important to know the percentage linear contraction, $L$. This is related to the volume contraction $V$, by the equation

$$\frac{V}{100} = \frac{3L}{100} - 3\left(\frac{L}{100}\right)^2 + \left(\frac{L}{100}\right)^3$$

For most purposes, it is convenient to ignore the smaller terms and take $L$ as a third of $V$. Also it is convenient to take $\rho = 2.5$, as this will be roughly true of many pottery materials. For some materials, however, such as alumina, a considerably higher value must be used.

From the above equations, it is possible to calculate $M$ and $L$ for a series of values of the unfired porosity $p$. The results of such calculations are shown on Fig. 4.4, for a range of moisture contents covering those normally used for plastic forming processes. The absolute position of any given whiteware body system is best determined by experiment. When this has been done, the chart can be used to predict the direction and likely magnitude of changes resulting from alteration of the packing arrangement.

For example, assume that a particular body system has an unfired porosity of 32 per cent, and it is found in practice that the optimum moisture content for a particular plastic making process is 22 per cent. From the chart **(Point A)** it can be seen that the wet-dry contraction is about 4.8 per cent. Incidentally, it may be noted here that skilled operators of traditional hand operations are particularly sensitive to the changes in workability owing to changes in moisture content, and usually require this to be correct within about ½ to 1 per cent. What degree of control might be required when fully automatic machines are used is a matter for speculation, but the requirements are likely to be at least as stringent. Now suppose that it is decided to make changes in the body composition such that the unfired porosity changes to 33 per cent, and it is then found that the optimum working moisture content is 24 per cent. The chart shows that the contraction is now about 5.9 per cent. **(Point B).**

Given information about the unfired porosity, this can thus be used, together with the working moisture content, to predict changes in wet-to-dry contraction with a degree of accuracy sufficient to serve for many practical situations.

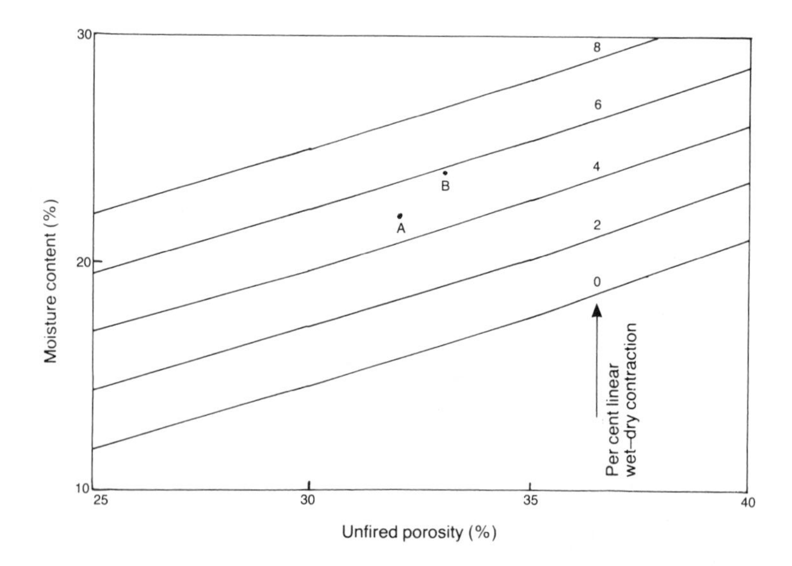


Fig. 4.4 – Wet–dry contraction relationships.

However, one or two qualifying observations must be made. It is found in practice that ball clay has special characteristics in relation to moisture content and workability. Bodies with higher ball clay content seem to need more water for a given degree of workability, and this results in a higher contraction than would be expected from the packing analysis. Also, in processes where an appreciable degree of pressure is applied to the body, as in some forms of extrusion, there is often a change in the porosity–water relationship, and hence in the contraction. Finally, it should be noted that in all the foregoing analysis, it has been assumed that the contraction is isometric. In practice, many forming processes result in an orientation of the clay particles in a particular direction. Thus, in an extruded bar, for example, the clay particles are aligned along the direction of extrusion, and the contraction is greater along the axis than across a diameter. So it is necessary to develop the particular relationships applying to the forming method in use.

## 4.4 UNFIRED STRENGTH

Unfired strength is of critical importance in relation to many forming and fettling processes, and an inadequate level of strength can result in serious manufacturing losses. Although many theories have been advanced to explain the surprisingly high strength of clay-based bodies when all the water has been removed, the explanations do not seem to be adequate. It is possible, however, to arrive at some useful practical conclusions by a simple consideration of the particle arrangement in the dry structure. Since the particles are in contact with each other, short-range van der Waals' forces will be operating at the points of contact, and these may be supplemented by the binding effect of any soluble matter left behind when the water has evaporated. In any case, it is reasonable to suppose that the strength will be related to the number of points of contact per unit volume, and this is very dependent on the packing of the particles, and hence can be related to the unfired porosity.

Consider the simple case of a cubic arrangement of spheres. If the spheres are of equal size, of diameter $D$, the porosity in this arrangement is 47.6 per cent. The space between eight spheres in contact can be occupied by a sphere of diameter $(\sqrt{3} - 1)D$; this reduces the porosity to 27.3 per cent, and increases the number of contacts from 12 to 20. The eight newly-created cavities can contain spheres of diameter $0.24D$; reducing the porosity to 21.4 per cent, and increasing the number of contacts to 44. And so on. As the packing improves, the porosity is reduced and the number of points of contact, which are an indication of the unfired strength, rapidly increase. Since the absolute size of the spheres also influences the number of points of contact, but not the porosity, the relationship needs to be represented by a family of curves of the shape shown in Fig. 4.5, each curve relating to a different degree of fineness. The decreasing particle size is indicated by an increasing number of unit cells per unit volume. On the same graph is shown a curve representing the measured values of the strength in a body system in which a wide range of unfired porosity was achieved by changing the size of both the clay and the filler, thus altering the packing from both ends of the system. It will be seen that the curve is of a reasonably similar shape to the theoretical curves, showing that large increases in unfired strength can be achieved by reducing particle size and improving packing. In fact, it is also found that there is good correlation between the strength values and the particle size of the constituent materials, as might be expected. Reducing the absolute size and at the same time improving the packing results in a much greater strength increase than is achieved by either change separately.

It seems clear, then, that consideration of particle size and packing arrangement will yield useful guiding information about unfired strength in any particular system. It should, however, be noted here that unfired strengths are sometimes found to be inadequate for a particular forming process, and have to be increased artificially by the addition of organic binders. In these cases, the effect of packing may well be masked.

## 4.5 PERMEABILITY AND CASTING RATE

Casting is a very important forming process, essential where irregular shapes are concerned. In this process, a fluid ceramic slip is placed in contact with a plaster mould; the suction of the mould extracts water from the slip, and a layer of solid cast is built up. The physical basis of this process is dealt with in more detail in Chapter 9, but at this point it is sufficient to note that the rate at which the cast thickness increases is strongly influenced by the permeability of the cast itself. The casting rate is an important factor in productivity, high casting rates

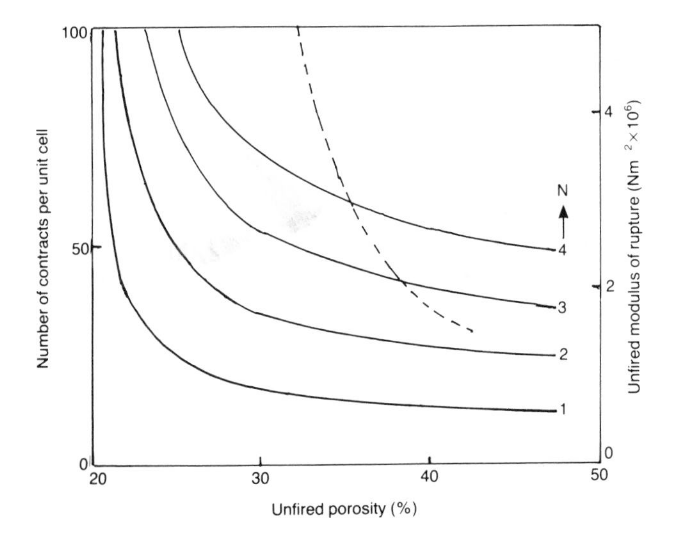


Fig. 4.5 – Relationship between unfired porosity and strength. —— Theoretical curves for cubic packing $N$ = number of unit cells per unit volume; ––– Measured values.

enabling a greater number of articles to be produced from a particular plant in a given time. It has to be said, however, that with very high casting rates other properties, such as the strength of the cast, may be adversely affected, so that an appropriate compromise is necessary.

Nevertheless, in situations where, for any reason, an increase in casting rate is desired, it is useful to know what effect changes in body packing are likely to have. In general terms, it might be expected that a lowering of porosity by improving the packing, or a reduction in pore size due to using finer materials would both reduce the permeability and slow down the casting rate. Before testing this hypothesis quantitatively, a qualifying factor may be noted here. In the casting process there is a strong tendency for the plate-like clay particles to be deposited with their long axis parallel to the mould surface, and this results in a lower permeability than would be expected from considerations assuming an isotropic distribution of particles.

The two important parameters to be considered are the volume porosity and the pore size. The volume porosity and the permeability are linked together, but not in a simple way. Much theoretical and experimental work has been done on this relationship, and many questions remain unanswered. However, an equation due to Kozeny has sound experimental support and is widely accepted. According to this equation, the permeability (and hence the casting rate) depends on $D^2$, where $D$ is the mean pore diameter, and on $p^3/(1-p)^2$, where $p$ is the volume porosity. The function $p^3/(1-p)^2$ is shown as curve D on Fig. 4.6, and it expresses the way in which casting rate might be expected to increase as more open body structures were used.

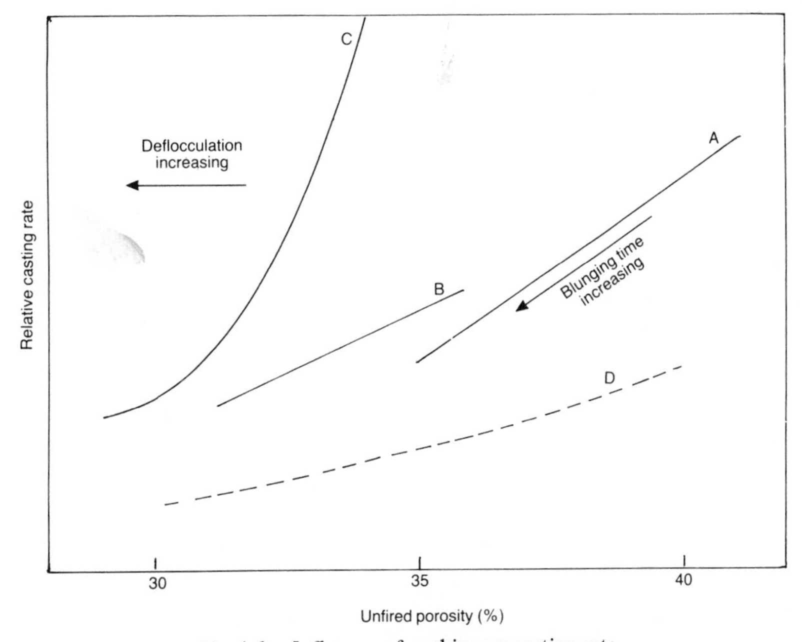


Fig. 4.6 – Influence of packing on casting rate.

Turning now to some practical cases in which the body packing is changed by various means, these are represented by the other three curves in Fig. 4.6. Curve A illustrates the effect of blunging. As the blunging time is extended the clay aggregates break down; the pore size is reduced and the packing is improved; and there is a predicted decrease in the casting rate. Curve B shows the result of substituting increasing percentages of ball clay for china clay. This progressively reduces the porosity and hence the casting rate. In the third case, curve C represents the results of a series of experiments with an earthenware body in which the concentration of the deflocculant was changed. Increasing the degree of deflocculation increased the density of the cast, reduced the porosity, and thus reduced the casting rate. In all these cases, the general rule appears to hold good. Improved packing, by whatever means achieved, has a marked effect in reducing the casting rate.

## 4.6 DRY-TO-FIRED CONTRACTION

In a previous section consideration was given to the effect of packing on the unfired porosity, the moisture content at forming, and the wet-to-dry contraction. In dealing with the further contraction that takes place on firing, it is necessary to make· the simplifing assumptions that the loss in weight takes place before the contraction commences, and that there is no change in the density of the particles owing to reaction in the firing. Subject to these, the following relationship may be established: if $p_u$ = percentage unfired porosity

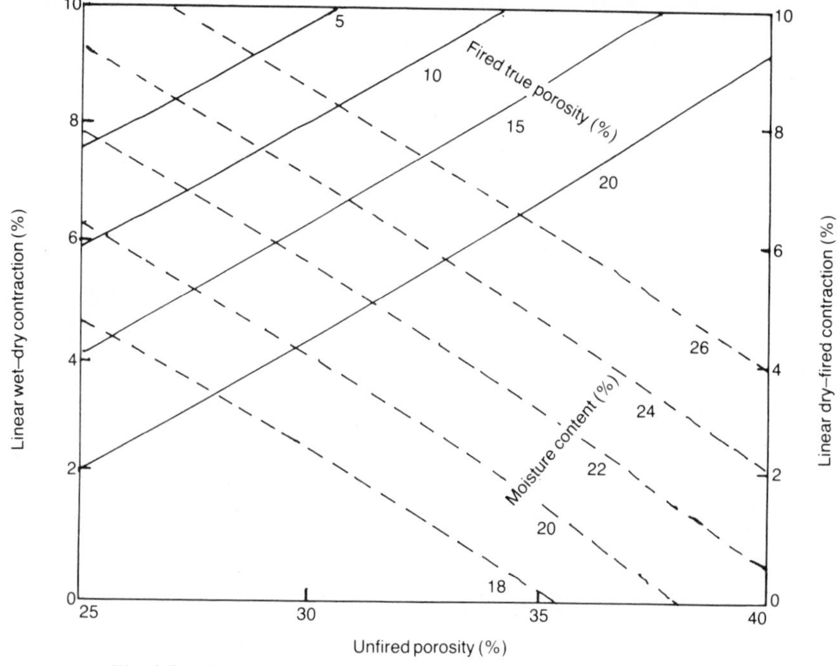


Fig. 4.7 – Contraction relationships. —— Dry-fired; – – – Wet-dry.

of dry body, and $p_f$ = percentage porosity of fired body, then the percentage dry-to-fired volume contraction is $100(p_u - p_f)/(100 - p_f)$. If the contraction is assumed to isometric, then the percentage linear dry-to-fired contraction is

$$100\left[1 - \sqrt[3]{\frac{100 - p_u}{100 - p_f}}\right]$$

It is thus possible to calculate the contraction from a knowledge of the unfired and fired porosities. The values so calculated are shown in Fig. 4.7, which also shows the data contained in Fig. 4.4. A chart of this kind, constructed for a particular body system, can be of great practical value. Starting from a known unfired porosity, that will be determined largely by the packing, and the moisture content used in the forming process, it is possible to read off the wet-to-dry and dry-to-fired contractions when the material is fired to a given true porosity. It should, perhaps, be noted at this point that the total contraction is not the arithmetical sum of these two values. Any change in packing will alter the unfired porosity, and probably also the optimum moisture content, and the chart indicates the likely contraction changes that will follow.

One further observation may be made. If the packing arrangement is altered by changes in particle size, there may well be a change in vitrification rate. In that case, it needs to be clear as to whether the firing is designed to be taken to a common temperature or to a common fired porosity, as the relative contractions will be different in the two cases. The general conclusion may be drawn that poor packing will usually result in high contraction.

## 4.7 FIRED STRENGTH

Fired strength will be dealt with in detail later, but, in so far as it is affected to some degree by the size of the constituent materials, it merits some consideration in the present context. The fact that the strength of fired whiteware bodies is a long way below the possible theoretical strength is related to the grain size of the crystalline particles present and to the degree of residual porosity. Grain size can be reduced by grinding, or other means. Good packing reduces the porosity, but even in the best conditions vitreous bodies can have a true porosity of the order of 5 per cent. In some practical cases, there is little reaction during firing and the crystalline filler materials, such as quartz and alumina, retain their identity in the fired body. In these cases, the grain size distribution of the starting materials will have a marked bearing on the fired strength. Earthenware would be a good example of such a system. In the case of bone china, where the fired constituents are quite different from the starting materials, the correlation would be expected to be much less marked. Nevertheless, the strength-porosity relationship is clearly applicable. It should be noted, of course, that the bonding between particles can be very different, so that materials of similar porosity can have very different strengths, if they have been subjected to different firing treatments.

## 4.8 CONCLUSIONS

In this chapter, it has been shown that the packing of the constituent particles in the body can have an important influence on many of the unfired and fired properties. The following practical points may be noted:

(1) Traditional whiteware bodies have not been designed in such a way as to have optimum packing arrangements.
(2) In a particular system, it is useful to set up a chart showing the variation of unfired porosity with packing. This helps in predicting the likely effect of changes in body composition.
(3) Wet-to-dry contraction can be related to the packing and the water content at forming. The effect on contraction, however, is largely compensated by changes in the amount of water needed to give optimum making conditions.
(4) Plasticity is influenced by packing, but is even more dependent on the intrinsic properties of clay (Moore and Lockett, 1966).
(5) Unfired strength is strongly dependent on packing.
(6) Permeability and casting rate are strongly dependent on packing.
(7) Most whiteware body systems conform reasonably closely to the generally accepted relationships between fired strength and porosity and grain size. Both depend to a considerable extent on the size distributions of the starting materials.

## REFERENCES

Dinsdale, A. and Wilkinson, W. T. (1966) *Tr. Brit. Cer. Soc.*, **65**, 391.
Furnas, C. C. (1928) *Bur. Mines. Rep. Invest.*, 2894, 7.
Moore, F. and Lockett, J. A. (1966) *Tr. Brit. Cer. Soc.*, **65**, 423.

# 5

# The role of water

Water plays a very important part in the processing and preparation of whiteware bodies. At first sight the addition and subsequent removal of large quantities of water might appear to be both cumbersome and thermally inefficient, but some of the functions it performs are indispensable. It makes it easier to remove impurities and to achieve a high degree of homogeneity in mixing. Easy flow facilitates pumping and transport. Most important of all, the appropriate water content provides for physical properties in the body suitable for particular forming techniques.

Before considering these different rheological states, it is necessary to establish certain definitions concerning the relationships between the weights and volumes of the solid and liquid parts, as there is frequent confusion about these. Although body recipes are normally expressed in terms of weight, some of the significant properties are best understood in terms of volume, so both need to be kept in mind.

We need first to define what is meant by moisture content, making a distinction between what are usually known as the dry basis and the wet basis. For a given mixture of water and solid, we use the following nomenclature.

$m_d$ = moisture content on dry basis
= weight of water as percentage of weight of solid.
$m_w$ = moisture content on wet basis
= weight of water as percentage of weight of water plus solid.
$\rho_s$ = specific gravity of solid, often taken as 2.5 for many ceramic materials, but very different for others.
$\rho$ = density of mix, in g/l.
$P$ = pint weight
= weight of a pint of the mix, usually expressed in ounces.
$p$ = porosity of dry body
= volume of pores as a percentage of total volume of solid plus pores.

It is nearly always better to express moisture contents as the wet basis, and whenever the term is used in this text without qualification the wet basis may be inferred. The use of pint weight has a long tradition in the industry as a means of expressing the density of the mix. Calculations of solid content are much simplified by the fact that a pint of water weighs approximately 20 ounces.

We may consider the formal relationships between the foregoing quantities. For a mix solid and water, and using simple algebra, it can easily be shown that

$$m_w = \frac{100 m_d}{100 + m_d}$$

$$m_d = \frac{100 m_w}{100 - m_w}$$

$$\rho = \frac{100 \rho_s}{100 + m_w(\rho_s - 1)}$$

$$P = 20\rho$$

and

$$m_w = \frac{100(\rho_s - \rho)}{\rho(\rho_s - 1)}$$

These relationships are shown graphically in Fig. 5.1, where the value of $\rho_s$ is taken as 2.5.

There is, of course, a continuous spectrum of possible water contents from 0 to 100, but not all mixes within this range have any practical value. Four ranges are of special interest to us.

Two regions on the right-hand side of the diagram relate to mixes in the liquid form. These frequently contain deflocculants in order to maximize the solids content at a given viscosity. The least viscous slips are suitable for mixing

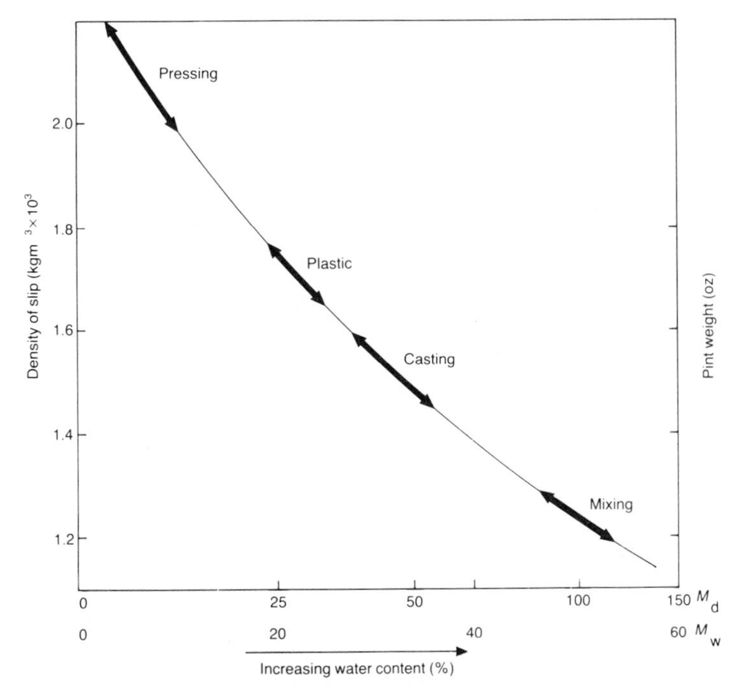


Fig. 5.1 – Forming processes and associated moisture contents.

and blunging, but higher solids contents are needed for casting. The casting process is very important industrially, as it is the only forming technique available for very complicated shapes; even where it is not the only option, it is often the most convenient. It is indispensable in the sanitary and tableware industries.

At lower moisture contents, the mix behaves more like a solid, but has the unique property of plasticity. This enables these mixes to be used for all plastic forming processes, essential in the tableware and electrical porcelain sections. At still lower moisture contents, the mix may properly be considered as a powder suitable for pressing. This forming technique is applicable to simple shapes, mainly in the tile industry, and also in electrical porcelain. It has the advantage of accurate dimension reproduction at high rates of output. It has increasingly come into prominence in recent years as a possible method for making flatware in the tableware section.

The manner in which these particular ranges of moisture content are achieved in practice is shown in Fig. 5.2. This flow sheet has been greatly simplified by omitting such stages as sieving and magnetting, which may not be necessary in any case where dry mixing is used with clean raw materials. Spray drying as a means of removing water has become very popular, mainly on account of its ease of operation and constancy of moisture content in the finished material. A high degree of deflocculation has to be achieved in the slip fed to the dryer. It is of interest to note that this dewatering technique can provide a route to the pro-

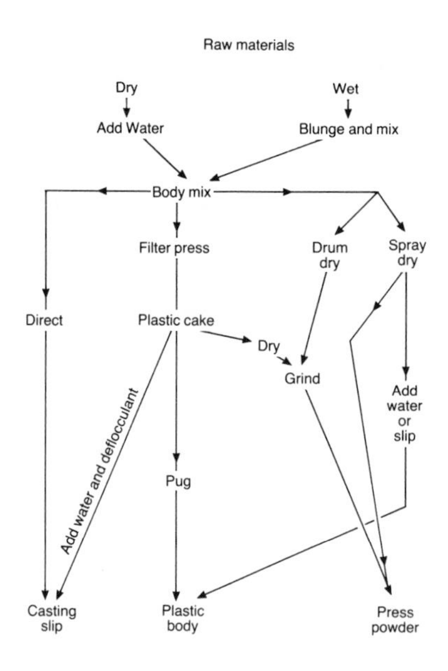


Fig. 5.2 – Simplified flow-chart for body preparation.

duction of plastic body as well as press powder. In the sanitary industry, the traditional method of preparing casting slip involved filter pressing, but improvements in sieving equipment have now made it possible to prepare the casting slip direct from the raw materials.

The properties of the mix are also important in connection with the processes which have to be performed after the article has been made. In order of decreasing moisture content, these include sponging, cutting, and fettling; drying; fettling and towing; and finally, at around 2 per cent, entry into the kiln.

Before turning to a consideration of the main forming techniques in detail it is useful to have a clear picture in mind of the relationship between the volumes of water and solid, and the volume of pores in the dry body. These are shown in the illustration in Fig. 5.3. It is assumed in all cases that the dry body after forming has a porosity of 35 per cent; for most industrial bodies it is in the range 30 to 40 per cent. It is seen that for powder pressing the pores are unfilled, and the function of the water is mainly to join the particles together. For plastic forming there has to be enough water to fill the pores and provide a surplus for lubrication. For casting, there is a larger surplus of water to provide liquid flow.

We proceed now to examine in detail the properties of the body mix for the three main forming techniques. An understanding of such elements as plasticity, strength, and slip rheology, will assist in elucidating the fundamental principles which underlie each process. We begin at the high water content end of the range.

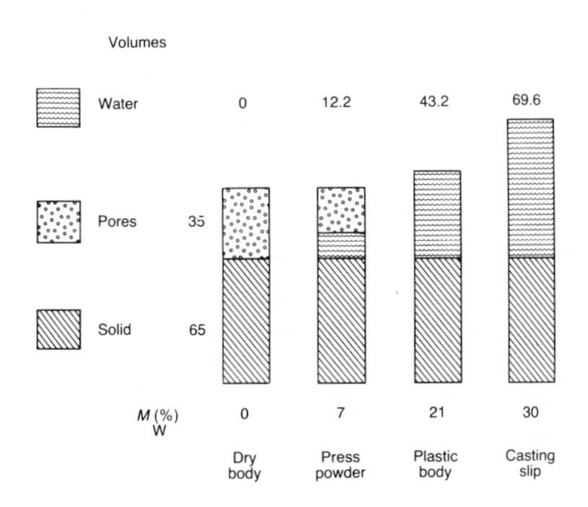


Fig. 5.3 – Water content in relation to porosity.

## 5.1 CASTING SLIPS

For those who are not familiar with it, the casting process might appear to be an unlikely method of forming a solid article. The process involves the use of a liquid suspension, or slip, which is brought into contact with a mould of the required shape. This mould is normally made of set Plaster of Paris, the fine pores of which extract some of the water from the slip so that a solid cast builds up on the surface of the mould. When a suitable thickness has been achieved the slip is poured off, and further drying enables the cast articles to be separated from the mould. In the case of open casting, the cast is formed on one surface only; in solid casting, on the other hand, the whole of the space between two mould surfaces is filled. The physical basis of the casting process is interesting, and is dealt with in detail in Chapter 9. Our concern at this stage, however, is the slip; to note those properties that are of significance to the process, and to consider how they may be optimized and controlled.

To do this we have to consider the rheological properties of clay-based suspensions. It may seem strange, at first sight, that rheology comes into it at all. The three major forming methods operate at very different levels of applied stress and rates of strain. In powder pressing, pressures of 1 ton/in$^2$ or more are commonly used. In plastic forming, pressures are lower than this, but the rate of strain is very high. Neither of these circumstances apply in the case of casting, where pressures are never more than atmospheric and rates of strain are very low. But it is found that a thorough understanding of the rheological properties of slips is vital to the effective operation and control of the process.

The two main requirements may first be stated. The first is that the slip must have a high solids content at an acceptable viscosity. Clearly there is a maximum permissible viscosity, or minimum fluidity, if the slip is to be transported, poured, and if it is to reach the recesses of complicated shapes. But this level of fluidity must not be achieved by reducing the solids content or the slip will not produce a solid cast in reasonable time. Solids contents of the order of 70 per cent are often required. The second requirement is that the cast should have the right properties. In short this means that it must be dense, firm, and strong enough to withstand handling, and that it should not crack or be so brittle that it cannot be cut or sponged.

In meeting both these requirements the manipulation of the rheological characteristics of the slip is necessary. The relevant parameters are density, viscosity and thixotropy. Before considering these in more detail, it should be mentioned that the rheological properties of slips derive mainly from the characteristics of clay, so that an understanding of clay suspensions is a necessary preliminary. Rheology is a difficult subject, mathematically difficult to handle and in terms of the chemistry only imperfectly understood. A detailed discussion of some of the conflicting theories would not be helpful in the present context, so we confine ourselves to general concepts, and verifiable experimental facts.

Kaolinite particles in a suspension of pure water, carry a negative charge, and there is thus a repulsive force between them, decreasing rapidly with the distance apart. Counterbalancing this, there are forces of attraction which also diminish, and even more rapidly, with distance. The balance of these forces can be altered by exchanging the ions on the clay surface. If sodium hydroxide (NaOH) is added to the solution, $Na^+$ ions begin to replace the hydrogen ions in the clay. At first this has little effect on the viscosity of the suspension, but at the concentration when all the hydrogen ions have been replaced, and there is thus NaOH in the liquid, there is a dramatic drop in the viscosity. The change may involve a factor of hundreds or even thousands, depending on the electroyte added. What has happened is that the balance of attractive and replusive forces has been radically altered, and the suspension is said to be deflocculated, i.e. the flocs or aggregates of particles have been dispersed. The rapid change in viscosity is shown qualitatively in Fig. 5.4(b). In practice, this change in viscosity can be brought about by a wide range of deflocculants, such as sodium carbonate, sodium silicate, mixtures of these, sodium hexametaphosphate (calgon), and others, all of which are much better than sodium hydroxide. Whatever deflocculating agent is used, the principle is the same; deflocculation enables casting slips to be made at much higher solids concentrations while maintaining the required level of fluidity. The viscosity of clay suspensions increases rapidly with solids content, as shown in Fig. 5.4(a). This diagram covers a low range of concentrations, up to about 20 per cent by weight. The straight line represents the well known Einstein equation. For clay slips, however, as shown by the curved line, the viscosity increases much more rapidly than linearly, so that in the absence of deflocculants the slip becomes much too viscous to be usable. Deflocculant additions of the order of 0.1 per cent bring the viscosity down dramatically, and enable the solids content to be increased to an acceptable level. The sort of change that occurs over the range of real casting slips is shown in Fig. 5.4(c) for a typical earthenware slip, the curve representing a constant value of fluidity.

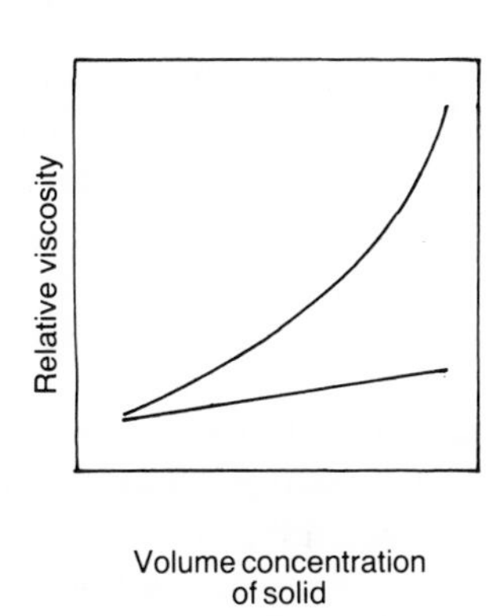


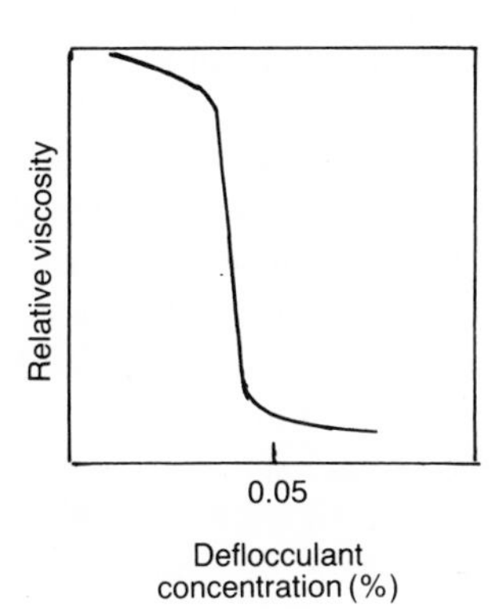

Fig. 5.4 – Viscosity relationships in slips.

Density of the slip, however, is not the only aspect of importance in practical casting operations. The condition of the deposited cast is also crucial, and the difference between flocculated and deflocculated slips is significant in this respect. If a slip is allowed to settle and form a sediment, it is found that the solid so formed is much more loosely packed in the case of the flocculated slip, and, as this open texture persists after drying, inadequate unfired strength would result.

Two other rheological properties need to be considered, yield point and thixotropy. The fluid characteristics of slips are often expressed in terms of the shearing stress associated with a given rate of shear. Moore (1959) has described the use of a torsion viscometer to provide information on these two variables for a number of ceramic slips. Typical curves are shown in Fig. 5.5. On the left of the diagram, the straight line A represents a pure viscous liquid, such as oil or syrup, which exhibits Newtonian behaviour. The rate of strain is proportional to the shearing stress, and the slope of the straight line is a measure of the reciprocal of the viscosity, which is independent of the rate of shear. On the right of the diagram, curves D and E relate to body slips, without deflocculant, at different densities. It will be noted that these curves do not pass through the origin, but cut the axis at a finite value of the shearing stress. This means that the slips exhibit a characteristic which, in the case of plastic materials, would be known as yield value. An initial stess has to be applied before any flow can take place, indicating that there are net attractive forces that have to be overcome. This is characteristic of flocculated slips in general. Curves B and C represent deflocculated body casting slips. Curve B is for bone china, and is unusual in that it is convex upwards. The deflocculation of china slips, on account of the high

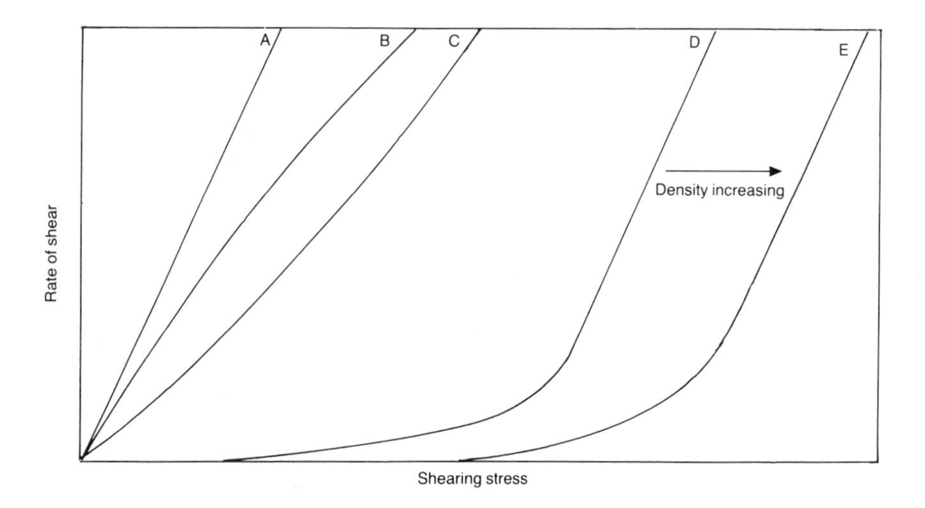


Fig. 5.5 – Stress–strain relationships for slips.

phosphate content, is a very complicated phenomenon, and is outside the scope of this book. Much more typical is the behaviour of an earthenware slip, shown in curve C. It is to be noted first of all that these curves pass through the origin. The absence of a yield value is typical of deflocculated slips, and it means that the repulsive forces outweigh the attractive forces at all inter-particle distances. Moreover, the gradient of the curve increases with increased rate of shearing, which indicates a lowering of the viscosity as the material is sheared. This is known as thixotropy, a word derived from the Greek words *thixis* = touching and *trope* = changing. When the stress is reduced, the viscosity increases again, and there is a dynamic equilibrium at any given rate of strain. The viscosity is also time-dependent in that when the stress is removed there is a return to a higher viscosity after a period of time. For practical purposes, a useful quantitative measure of the thixotropy is the change in viscosity after an arbitrary length of time.

The mechanism of thixotropy is not completely understood, but it is generally felt to be attributable to the presence of aggregates in the clay-based material, which persist even when it is deflocculated. Kaolinite particles, although carrying negative charges overall, are plate-like and can have very different charges on the edge and face. Edges may sometimes be attached to faces, and there is a range of stacking possibilities. The two structures shown in Fig. 5.6 are sometimes referred to as 'house of cards' and 'pack of cards'. It is postulated that when a material is sheared the structure is broken down, lowering the viscosity, but when the stress is removed, the structure would reassemble.

Whatever the reason, the phenomenon is of great benefit to the casting process, as it tends to assist in the building up of the cast. Casting slips operate best with a degree of thixotropy, though too much can be harmful. It is important to control both viscosity and thixotropy in order to obtain optimum casting condition, and thixotropy as well as viscosity is a function of slip density and deflocculant concentration.

This leads us to consider the question of casting slip control, the basic

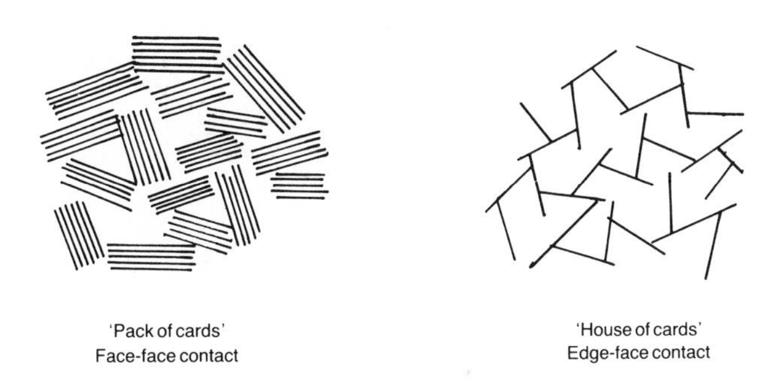


Fig. 5.6 – Possible arrangements of kaolinite particles in casting slips

principles of which have been clearly stated by Moore (1959). The relevant slip properties that need to be measured are density, fluidity, and thixotropy. Density presents no problem; the traditional method is simply to take the weight of a pint of slip. The other two present very real difficulties. There is a wide range of possible techniques for measuring viscosity, or fluidity, many of them being sufficiently sophisticated to give values for different rates of shear. However, for the requirements of routine factory control it has been found to be adequate to use the so-called torsion viscometer. It is virtually impossible to derive absolute values of viscosity from this instrument, but the so-called fluidity values are sufficiently indicative of the state of a slip for most control purposes. The apparatus consists of a cylindrical bob, supported by a thin torsion wire and allowed to rotate in a cylinder containing the slip. The cylinder is rotated backwards, against the torsion of the wire, through an angle of 360°. It is then released and the angle through which it travels past the zero equilibrium position is taken as a measure of the fluidity, generally quoted in degrees overswing. This reading is taken on a freshly stirred slip, and a second reading is taken at an arbitrary fixed time, usually one minute after the first. If the slip is thixotropic it will have thickened up slightly during this time, and the second reading will be lower than the first. The difference between the two reading is taken as a measure of the thixotropy.

There is no general formula for the best values of these parameters to produce good casting practice, as they vary markedly from one body to another. In these circumstances, it is necessary to measure the parameters for a particular body system, by varying the density and deflocculant additions and noting the changes that occur. When these measurements have been obtained, they can be plotted on what is often called a characteristic curve, such as is shown in Fig. 5.7 for an earthenware casting slip.

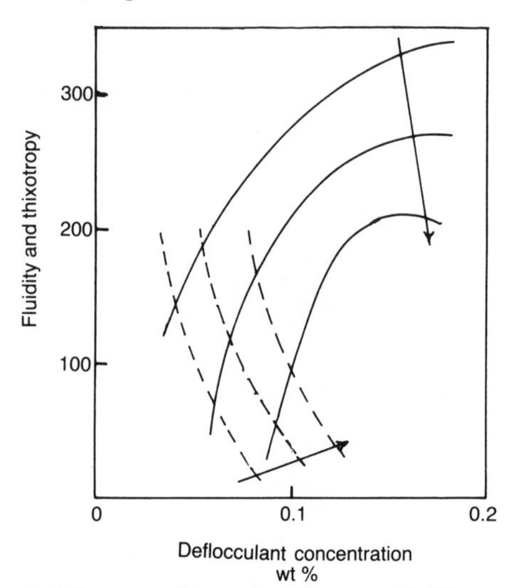


Fig. 5.7 – Characteristic curves for casting slip. —— Fluidity; – – – Thixotropy; ——→ Slip density increasing.

In any system there are, on the face of it, three variable quantities, namely body, water and deflocculant, but in fact these reduce to the two latter. It should be noted, however, that the variables are not independent. Adding water, for example, reduces the density, increases the fluidity, and decreases the thixotropy. Adding deflocculant in the form of a solution also increases the water content, and so on. Thus it is that the adjustment of the slip properties on the factory is a matter of successive approximation, using the characteristic curve as a guide. Looking at Fig. 5.7, we see that, as we have previously noted, there is a rapid increase in fluidity for small deflocculant additions, coming eventually to a maximum. Casting slips are generally used slightly to the left of this maximum, with a fluidity around 250 to 300°. Fluidity decreases as the pint weight increases. Thixotropy, which is usually somewhere in the region of 20-50° increases with pint weight, but decreases with increasing deflocculant concentration. The main use of this type of control system is to ensure continuity of slip properties from day to day, and in this connection it must be mentioned that deflocculated slip properties change with time, so ageing has to be taken into account.

However, as well as constancy it is necessary to establish suitability. The absolute values of the properties appropriate to a particular casting function have to be found, and this can only be done by experiment. We can only indicate here the practical results that are of most significance, and, in general terms, how they may be expected to vary from one slip condition to another. Most important of all to the manufacturer is the casting rate, and production demands require that this should be as high as possible, consistent with other aspects being acceptable. These other aspects are equally important, and they include the consistency of the cast and a range of possible production losses. Beginning with the constitution of the body itself, our previous consideration of packing density would lead us to the conclusion that a loosely packed structure would increase the casting rate; on the other hand, it makes for a porous cast with inadequate strength. The way to high casting rates is through high solids content, with the necessary level of fluidity achieved by deflocculation. A moderate degree of thixotropy helps to strengthen the cast and to prevent brittleness. When the thixotropy is too high, flabby casts will result, and there may be problems with draining or drying. Shrinkage, warpage, cracking, and wreathing are other aspects that need to be considered. The purpose of the casting slip control system is to maintain the optimum condition when it has been identified and achieved. The necessity for frequent and detailed control derives from possible variations in raw materials, moisture contents, salt impurities, recirculated scrap additions, and other unforeseen changes. Lack of control can be very costly in terms of production loss.

Much is still to be learned about the rheological properties and the casting characteristics of slips. What has been described here outlines the basis of understanding for the relatively simple clay-flux-quartz body system. Other systems, such as bone china, behave differently in a number of important respects, and control systems do not carry over without modification from one system to another. The casting of non-clay systems is too specialized a subject to be dealt with in the present context.

We can sum up this section by noting that in the fluid region of the body: water mix, the rheological properties of the slip hold the clue to the practical casting behaviour. They can be measured fairly simply and, rightly understood, they can provide the basis for effective factory production control.

## 5.2 PLASTIC BODIES

Ceramic forming first began when man discovered the fascinating properties of plastic clay. Plasticity is a phenomenon that occurs in a number of solid materials. In covalent bonding, the atoms are not free to move easily and the material is thus brittle, strong and non-plastic. In ionic crystals, plasticity can occur at high levels of purity, but when imperfections are present movement is restricted. Metals display plastic flow to a marked degree. We are not here concerned, however, with the properties of single crystals, but with a mixture of body materials and water. As the moisture content of the mix is reduced in Fig. 5.1 we move from casting slips into a region where the body is sticky, and is of no use for forming purposes. At lower moisture contents we come into the plastic range, where some very important forming techniques operate. These include extrusion, jiggering of flat and hollow tableware, and throwing. The body is said to be plastic, meaning that it is capable of being moulded into a prescribed shape. The requirements may be stated in qualitative terms, as follows:

(1) The material must deform easily under an imposed stress without rupture.
(2) It must maintain the imposed shape after the stress has been removed, and so must not deform under the action of gravity or subsequent handling. This means that it must have an appreciable yield stress.

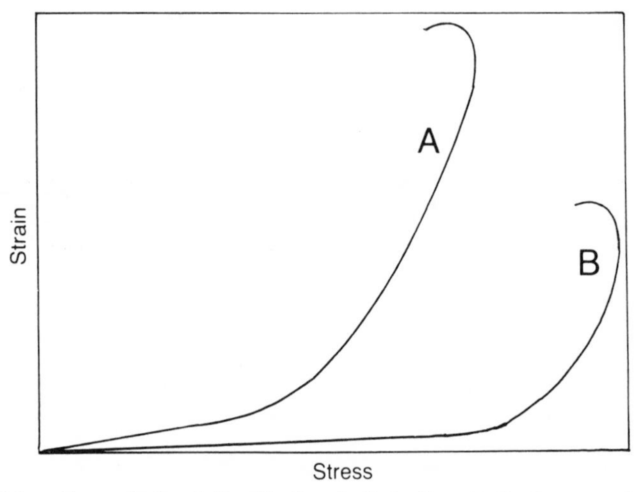


Fig. 5.8 – Stress–strain relationship for plastic bodies. A, Earthenware; B, bone china.

How these requirement are met is best understood by considering the rheological properties of the body, which are similar in some respects to those of slips, except that elastic deformation enters as well as viscous flow. There are many ways in which measurements can be made on the stress–strain relationships; by rotation of hollow or solid cylinders, cyclic torsion, extrusion under pressure, and so on. The practical implications of plasticity in bodies have been discussed by Bloor (1957, 1959) and by Baudran and Deplus (1959). The quantitative results are not easy to correlate, but we can see clearly the kind of picture that emerges.

Fig. 5.8 shows the general form of the stress–strain curve. The main characteristics are that

(1) Initially there is a small elastic deformation, represented by a straight line.
(2) When the yield stress is reached the material begins to deform much more easily.
(3) After a certain amount of deformation has taken place, rupture occurs.

Similar characteristics emerge, if we plot the stress against the rate of strain, as shown in Fig. 5.9.

This rheological behaviour derives from the special nature of the clay particle. Because of its surface properties it adsorbs water readily, and it is known that this water has two effects; it holds the particles together by surface tension forces, but when these have been overcome it allows flow to take place easily. The flow is much facilitated by the plate-like nature of the particles which enables them to move easily with respect to each other. When the imposed stress is high enough the particles become separated and the attractive forces can

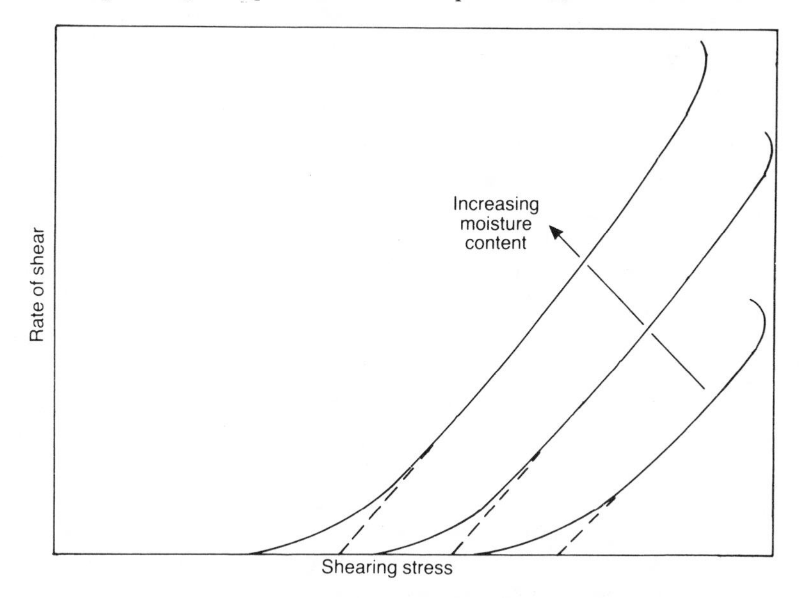


Fig. 5.9 – Stress-rate of strain relationship for plastic bodies.

no longer hold them together; rupture occurs.

We need to consider in more detail the practical implications of the shape of the curves in Fig. 5.9, with respect to different forming techniques, and then what can be done to the body mix to modify the properties. The features to be noted are

(1) The yield stress, which is the point at which easy flow sets in. This is often indistinct, but an arbitrary value can be obtained by extrapolating the straight part of the curve back to zero rate of shear. It will be seen that this yield stress increases with decreasing moisture content.
(2) A section of approximately viscous flow. This is known as the deformation range, and it is the range over which it is possible to carry out plastic forming. The length of this usable range increases with increasing moisture content.
(3) At the end of the deformation range, a point at which the rate of shear suddenly increases without any increase of stress, and rupture occurs. With increasing moisture content, the stress at which rupture occurs decreases.

In making plates on a semi-automatic jigger, the plate is supported on the mould so there is no need for a high yield value. Since the yield value decreases at higher moisture contents, this means that the clay can be used in a wetter condition. This extends the deformation range, since the deformation involved in this process is very large. Even here the yield value is of quite a different order of magnitude to that associated with casting slips (about $10 \times 10^6$ $Nm^{-2}$ or more, compared with $0.1 \times 10^6$ $Nm^{-2}$). A similar long deformation range is needed for throwing. Roller-head making can be carried out with much stiffer material and extrusion of large articles stiffer still; a high yield stress may well be needed in the latter case, as the products can be very large and heavy. How can the required properties be achieved? We need to distinguish between the changes that can be brought about by changing the moisture content, as have already been noted, and those that arise from intrinsic properties in the body material. A look at Fig. 5.8 gives us a clue. The two curves shown there illustrate the difference between earthenware and bone china. The earthenware body has a lower yield stress and longer deformation range than the bone china. The practising potter would immediately recognize this, and would call one 'fat' and the other 'short'. Plastic forming is much easier with earthenware than with bone china. The significant difference is the quantity and nature of the clay content. The plastic behaviour of bodies derives almost entirely from the clay, and the most important parameter is the fineness. Fine-grained clays have much more affinity for water, and therefore exhibit higher plasticity. So ball clays are more plastic than china clays. Thus, earthenware with ball clay and a higher total clay content is more plastic than bone china. Very fine clays, such as bentonite, can increase the plasticity still further.

Some of the other relevant factors may be noted. Good mixing practice and vacuum pugging, by evenly distributing the moisture and reducing the bubble content, improve plasticity. Good packing giving lower unfired porosity gives better plasticity. In some contradiction to this is the well known practice of adding acid, or some other flocculant, to a body to improve its workability. But the most important thing is to choose the type and quantity of clay to meet the demands of the forming process used, and then to adjust the moisture content to optimize the workability.

It will occur to the reader that nothing has yet been said about defining, let alone measuring, the so-called plasticity index. The reason is that both these are impossible. Many proposals have been made, using various arrangements for applying the stress and measuring the resulting deformation, for calculating a number that would represent a plasticity index that is an intrinsic property of the body and is independent of the moisture content. None is wholly satisfactory, in the sense that it is not possible to achieve an infallible correlation between the index and the known practical behaviour of the body. A very interesting approach has been proposed by Astbury (1963) involving the analysis of the hysteresis loop obtained by cyclic shearing of a cylinder in torsion.

The search will no doubt continue, but the phenomenon is so complex that a simple solution is not likely to be forthcoming. In the meantime any method of measurement that will provide the required stress–strain curves will give the necessary control guidance, along the lines discussed above.

## 5.3 PRESS POWDERS

Semi-dry pressing is the forming technique widely used in the manufacture of tiles and some electrical porcelain components, and is increasingly coming into consideration as a possible method for making plates in the tableware industry. It is particularly effective where simple geometrical shapes are concerned, and it offers great scope for automation and high rates of production, combined with dimensional accuracy.

We are concerned here with moisture contents in the range 5 to 10 per cent, and, as we see from Fig. 5.3, the pore volume is only about half-filled by the water. The function of the water is to increase cohesion by surface tension forces, and by the adhesive bridges built by the residual solids content of the water after drying. The rheological properties of the body mix are not significant here, in the sense that they were in the case of plastic body or casting slips. There is, however, some need to consider the bulk flow properties of the press powder in relation to flow to the die, and distribution in the die.

The compaction of powders is a complex subject, and in some aspects difficult to understand. Pressing powders that are completely dry is very difficult, and the pressing of a dried body mix would be unsatisfactory. It is necessary first to convert the body into granules which flow easily into the die and which crush when subjected to a pressure of about $30 \times 10^6$ $Nm^{-2}$ (about 2 ton/in$^2$) to give a strong, densified, compact. It is of interest to consider more closely the nature of these granules. The granules normally have a size ranging from 100 to 500 $\mu$m, so that each granule contains a very large number of constituent particles.

If we take spheres of 1 $\mu$m diameter to represent clay, and 10 $\mu$m to represent flux and filler, then it is easy to calculate that a spherical granule with a diameter of 100 $\mu$m will contain of the order of 1000 particles of flux or filler and 1 000 000 particles of clay. This means that, if the mixing has been thoroughly carried out, each granule represents in microcosm the proportions and properties of the mix as a whole, and if this is so, uniformity of composition throughout the pressed compact is greatly assisted.

It follows that the method of producing the granules, and the characteristics of granules, needs careful consideration.

The wetting up of a dry powder is a very difficult operation, although it is sometimes achieved satsifactorily. The mainstream methods of producing press powders, therefore, start from wet mixing, which gives the required degree of dispersion and homogeneity. The subsequent drying down may be carried out in many different ways, of which only those widely in use are described here.

The most traditional method is to filter press the body mix, and to further dry the filter cakes to a moisture content suitable for comminution and sieving. An alternative route is to extrude the filter cake material in the form of 'spaghetti' filaments to make drying easier. The moisture content at which grinding takes place may be slightly lower than that needed for pressing, so that a small degree of wetting up may be needed. Another technique widely used for the drying of the body slip is that known as drum drying. In this method, a heated drum in contact with the slip is rotated at such a speed that it picks up a thin layer of slip, which dries sufficiently for it to be scraped off by blades sited further round the circumference of the drum. The grains produced by these methods are all similar and are usually called angular. They are irregular in shape, having sharp edges and corners, as shown in Fig. 5.10. They are also rough on the surface.

By contrast, a dewatering technique that has become increasingly popular in recent years is spray drying, commonly used in other industries (Fig. 5.11). In this process the liquid slip is atomized and sprayed into a stream of hot gases, where evaporation from the droplets produces a granular material in a single, easily controlled operation. This method has the advantage of continuous operation, and a high degree of consistency, especially of moisture content, in the product. It may seem surprising on first sight that such high rates of evaporation

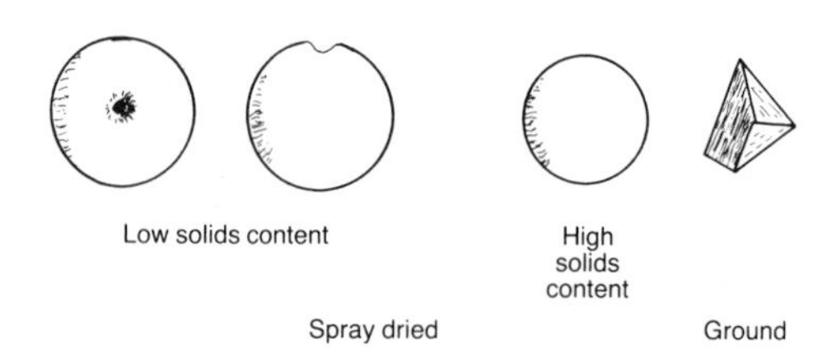


Fig. 5.10 – Granule shapes.

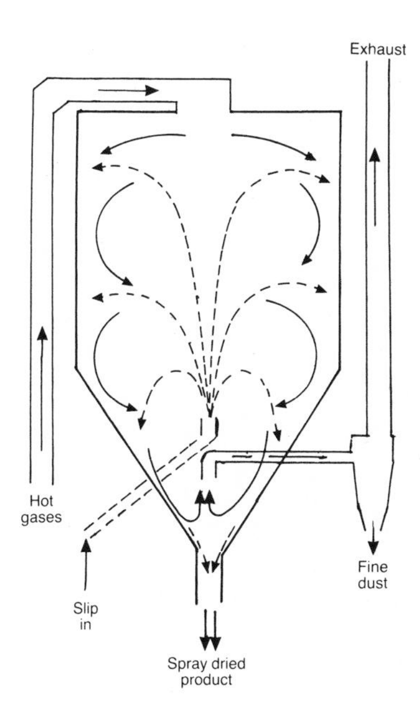


Fig. 5.11 – A spray dryer.

are possible. However, it can be calculated that a drop of water of 100 $\mu$m in diameter will evaporate completely in 0.1 s at a temperature of 400° C. Body slip, of course, will evaporate more slowly, and generally the surface rate of evaporation will exceed the rate at which water can flow through the rest of the particle. Thus, instead of solid sphere resulting, the grain is hollow, as shown in Fig. 5.10; a shape often referred to as 'doughnut'. If the solids content of the slip is sufficiently increased by deflocculation, solid spheres can be made. The spherical nature of these granules makes them behave quite differently from the angular grains produced by the other techniques, and these differences may be summarized as follows.

*Flow.* The spray dried granuels flow much more easily than angular. This is an advantage in transport to the die, but on the other hand the angular grains are better able to retain an imposed shape. The flow characteristics can be readily ascertained by tilting a heap of powder and observing the angle of flow, or by observing the angle of rest in a heap.
*Densification.* Angular grains have a wider size range and thus pack more compactly. Hollow spheres may give a more porous compact.
*Sticking to die.* A point to note here is that when spray drying is used the deflocculant is retained in the grain, and may be highly concentrated at the surface. This may have an influence on the tendency to stick to the die. Sticking to the die may also occur if the water content of the granule is too high.

*Surface finish.* Smoother surface finishes can be obtained with spray dried material, though it must be noted that the significance of this varies greatly from one product to another, being particularly high in the case of tableware.

So there are advantages and disadvantages associated with each method of granulation.

What aspects of body preparation are important in the technology of powder pressing? The body ingredients will be chosen mainly to give the required properties in the fired state. However, since pressed compacts are not characterized by high unfired strength, an appreciable clay content is necessary to help in this respect by increasing the number of contacts rather than by conferring any special rheological features, such as plasticity. The control of moisture content is critical. Increasing moisture content over the relevant range increases the unfired strength markedly; from 6 to 8 per cent may bring about a strength increase of 20 per cent. On the other hand, too much moisture diminishes the bulk flow properties and decreases the air permeability, so that the evacuation of air is more difficult. When the optimum moisture content has been identified, it needs to be controlled within very close limits for successful continuous production to be maintained.

The considerations outlined in this chapter confirm that the role of water in whiteware bodies is as important as that of the solid components. Its absolute value can determine the properties needed for a particular forming technique, and its control is important in all of them.

## REFERENCES

Astbury, N. F. (1963) *Tr. Brit. Cer. Soc.*, **62**, 1.

Baudran, A. and Deplus, C. (1959) *Tr. Brit. Cer. Soc.*, **58**, 454.

Bloor, E. C. (1957) *Tr. Brit. Cer. Soc.*, **56**, 423.

Bloor, E. C. (1959) *Tr. Brit. Cer. Soc.*, **58**, 429.

Moore, F. (1959) *Tr. Brit. Cer. Soc.*, **58**, 470.

Moore, F. (1959) *The A.T. Green Book, 1959,* 201 (British Ceramic Research Association).

# 6

# Semi-dry pressing

Formation of relatively simple shapes by pressing is widely used in ceramic production. It can be carried out with dry powders, by using added binders and lubricants, or with plastic body, but the most widely used method employs powder at an intermediate moisture content, of the order of 8 per cent. It is this range, generally known as semi-dry pressing with which we are concerned in this chapter.

The method has many important advantages. It lends itself to the rapid and continuous production of dimensionally accurate articles, and can readily be fully automated. Its principal use is in the tile and electrical porcelain sections, but its potential in the forming of flat tableware items, such as plates, is the subject of much current investigation and development.

Presses are of various types, but may conveniently be divided into hydraulic and mechanical. Hydraulic presses offer a smooth continuous loading; mechanical presses can give more of an impact type of pressing. A double press action is often claimed to give better results as the initial press facilitates the removal of air. Modern tile presses can press a number of tiles in one stroke, giving output rates of more than 200 tiles per minute. The whole process from powder transport, through die filling and pressing, to ejection and stacking of the tiles, can be fully automated (Fig. 6.1).

The differences between angular and spray dried dust are significant and have been discussed in Chapter 5. However, the general features of the pressing process apply equally to both. We consider now what happens during the compaction process, and what is known of the physical properties of the pressed compact.

## 6.1 COMPACTION

A number of variables can be identified as having an important bearing on the

Fig. 6.1 – Automatic press unit, producing and stacking 234 4¼-inch wall tiles per minute. (Courtesy H and R Johnson Tiles.)

compaction process. At the beginning, the moisture content of the powder has and influence on the loose fill density in the die box. When the powder is too dry, there is not enough moisture to exert adequate surface tension forces to draw the particles together; when it is too wet, sticking occurs and flow into the die is restricted. An optimum condition is in the range from 6 to 10 per cent. It should be noted here that granulation gives an unpressed density appreciably higher than that obtaining in a powder of the same material.

When the pressure is applied the plunger may travel a distance of several inches. During the first part of this stroke the pressure is very low. It is not sufficient to crush the granules, but there is a rapid rearrangement of granules into the available spaces, on a local basis with little lateral flow in a macroscopic sense. However, the rearrangement brings about a rapid initial increase in bulk density. During the remainder of the stroke, the pressure increases to a value that is high enough to crush some of the granules, thus further increasing the density. Further increase in pressure results in elastic compression. Little advantage is gained by increasing the pressure beyond this point, so that practical industrial pressures are of the order of $30 \times 10^6$ Nm$^{-2}$, or 2 tons/in$^2$. At this pressure the bulk density is about 70 per cent of theoretical, slightly higher than that achieved in plastic making. The corresponding unfired porosity can be calculated from the formula

$$\text{porosity} = 1 - \frac{\text{bulk density}}{\text{specific gravity of solid}} \times 100$$

The final porosity is thus around 30 per cent.

An interesting feature of the final elastic compression is that it is reversible, so that when the compact is removed from the die, it expands by about ½ per cent in a linear dimension. What is not well understood is that this elastic recovery may take place over an appreciable period of time, sometimes lasting more than 24 hours.

A parameter that is of vital significance in die design, especially where there is a variation from a regular geometrical shape, is the compaction ratio. This is defined as the ratio of the initial to the final depth, and it is usually of the order of 2, or slightly more. Since the process of compaction is a complex one, involving both rearrangement and crushing of granules, the relationship between compaction ratio and bulk density or porosity is not simple.

Many attempts have been made to deduce formulae relating bulk density to pressure; the curves can sometimes be fitted by a number of exponential terms, but the curve is rarely of a simple exponential form. Its shape varies with material and size distribution. However, a constant feature is the rapid compaction at the beginning, and the low return for increased pressure towards the end.

One of the critical variables in the process is the moisture content of the powder. From the point of view of bulk density there is an optimum moisture content around 8 per cent. At lower values there is not enough water to facilitate granule rearrangement, while at higher values the water appears to keep the granules apart.

## 6.2 PROPERTIES OF THE COMPACT

### 6.2.1 Strength

Unfired strength is a very important property; if it is inadequate, problems arise with handling and cracking. The effect of pressing conditions, and the practical consequences, have been studied for an earthenware tile by West (1955). The complex of forces that provide the strength is not fully understood. It is clear, however, that surface tension forces due to the water film, allied to the applied pressure, succeed in bringing the particles into sufficicently close contact for van der Waals' forces to be operative. At the relevant particle sizes and inter-particle separation distances, it can be shown that these forces are of the right order to account for the measured strength.

It is found that the modulus of rupture of a tile as pressed varies with the moisture content of the press powder in the manner shown by curve **A** in Fig. 6.2, where it is again seen that an optimum value is in the region of 10 per cent.

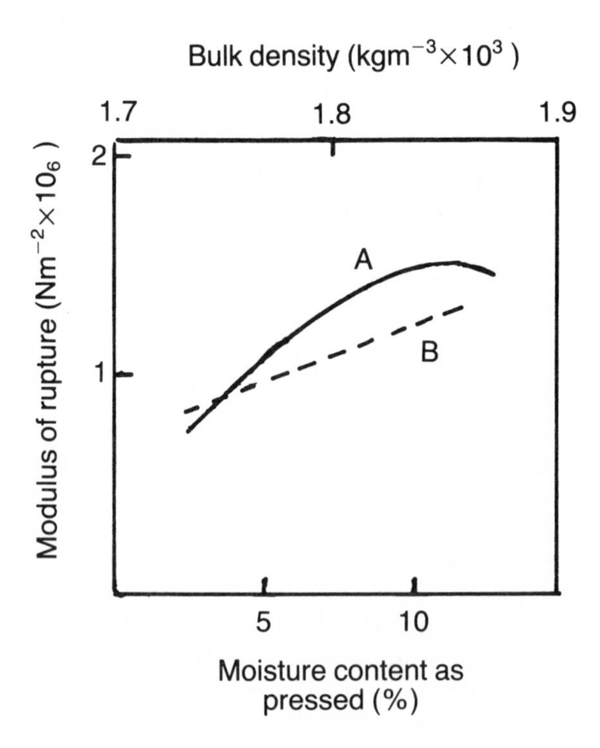


Fig. 6.2 – Effect of pressing moisture content on the strength of a pressed tile.

The relationship between modulus of rupture and bulk density is also shown by curve B in Fig. 6.2. The generally accepted connection between porosity and strength is not strictly followed, because of the complex nature of the compaction process. However, the strength does increase as the porosity is reduced, and it often turns out that the strength increases in an approximately linear manner with the bulk density, as shown on the graph. So far reference has been made only to the strength of the compact as pressed. In common with most whiteware bodies, however formed, there is a substantial increase in strength when the moisture is removed; as shown in Fig. 6.3 (Holdridge, 1952). As the compact is dried there is very little change in strength, and negligible contraction, down to a moisture content of about 2 per cent. Thereafter, the strength increases rapidly until it may be doubled when the completely dry condition is reached.

### 6.2.2 Density variations

In the above considerations, we have been concerned only with the bulk properties of the compact as a whole. But the compact is by no means homogeneous, and the variation in pressed density, together with the variation in other properties that follow from these, are important in some circumstances (Stuijts and Oudemans, 1965).

A primary cause is friction in the die. The powder is not fluid, and whilst

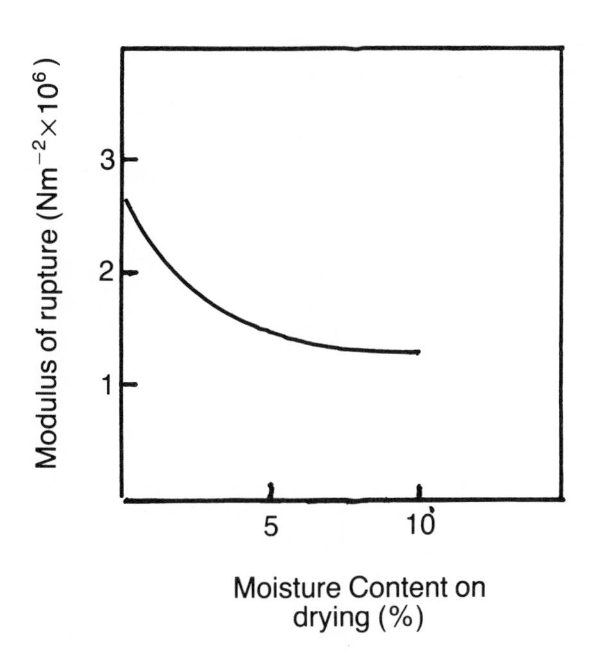


Fig. 6.3 – Effect of drying moisture content on the strength of a pressed tile.

there is some local lateral migration, it is not great. Friction at the walls of the die means that material does not compress as easily there; moreover, this friction is manifest to a reducing degree at distances from the wall. In addition to this there is a pressure drop from the face of the plunger vertically down through the powder. The stress distribution is very complex, and the density variations depend on the nature and size distribution of the material, as well as on the pressure. Nevertheless, there are some features that are generally present. Fig. 6.4 shows the pattern of density distribution across a vertical diametral plane in a cylinder pressed from the top only. A regular feature is the high density in the top corners and the low density in the bottom corners. There is usually also a low density area on the centre line just below the top surface. Sometimes there is a dense area on the central axis below the halfway line. These vertical variations in density are of no significance in the case of pressed tiles, where the vertical dimension is very small. For electrical porcelain and special ceramic components they may be extremely important. The variations can be reduced by a number of technical devices; these include vibration, both during the filling of the die and during pressing; pressing from both top and bottom; and the use of lubricants.

What may be significant in the case of tiles is the variation along the face. The density varies along the edge of the tile and along the centre line of the face. Along the edge there is slight loss in density at the centre compared with the corners. Along the centre line there is further slight drop to the centre of the tile. Variations at these levels only become important in so far as they affect glaze take up, surface finish, and fired properties. In cases where they are exaggerated by asymmetrical die filling or bad press adjustment they can give rise to the phenomenon of wedging. Differential contraction can cause pairs of opposite sides to be non-parallel. Cubbon (1982) has described some of the problems that may arise with less simple shapes, such as dinner plates.

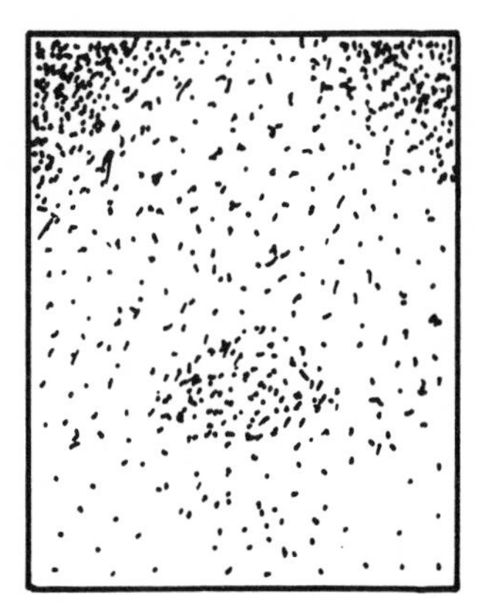

Fig. 6.4 – Density variation in a diametral section of a pressed cylinder.

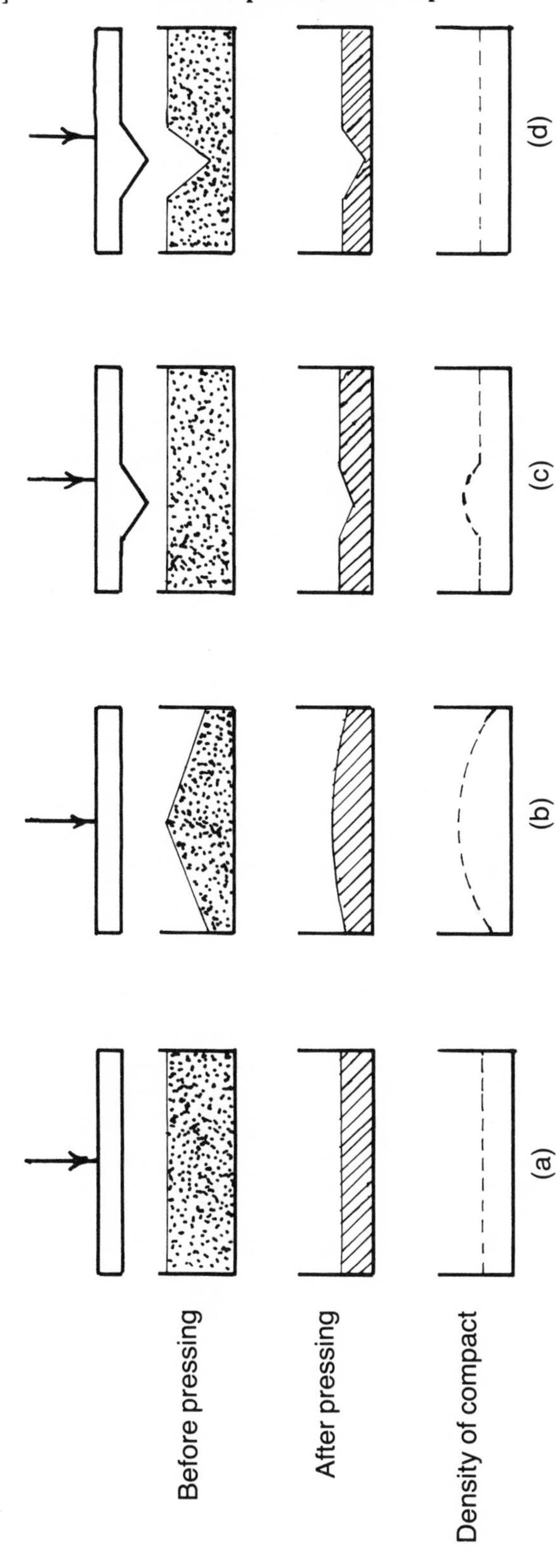


Fig. 6.5 – Influence of die shape and powder profile on pressed density.

Apart from irregular die filling and friction losses, density variations can arise from the difficulty of accommodating shapes which differ from simple geometrical profiles. The problem is to design a die shape, and a powder configuration, to give uniform density at all points in the article. We can see the elements of the problem by considering Fig. 6.5(a) which shows the case of a bed of uniform thickness subjected to such a pressure as to produce a compaction ratio of 2. The density of the compact is constant across the die, ignoring friction losses. If, however, the powder bed is not uniform in thickness, as shown in Fig. 6.5(b), the pressure will be greater at the centre than at the edges, and the compaction ratio and density will also vary accordingly. The thickness of the pressed compact will not be quite uniform, and the density at the centre will be higher than that at the edge. If now the profile of the plunger is not linear, as shown in Fig. 6.5(c), and if it is applied to a bed of uniform thickness, the shape of the resulting indentation will be different from the shape of the die, and there will be a density variation in the compact. This can be overcome by preforming the powder into a shape related to the die shape by a factor equal to the compaction ratio, giving the required shape in the compact, and a uniform density, as shown in Fig. 6.5(d).

This is not easy to achieve in practice. A section through a dinner plate shows appreciable variations in thickness, notably at the shoulder and abruptly at the foot. Making the die of a suitable shape, and spreading the powder in a correct profile, are very difficult operations, and complete homogeneity is almost impossible to achieve. The problem is also aggravated by having to press against a sloping side, such as the rim of a plate. If the angle of slope is too great it is not possible to profile the powder, and, in any case, the effective pressure varies with the cosine of the angle.

One method of minimizing the density variations is to use some modification of the so-called isostatic pressing technique (Fig. 6.6). In this method the powder is enclosed in a flexible membrane and pressure applied unformly in all directions. In the case of plate pressing the effect can be achieved by the use of a flexible membrane or a flexible coating on the die. Variations are thereby reduced, but dimensional accuracy is inevitably forfeited, so that a choice has to be made between dimensional accuracy and density variation. The significance of inhomogeneity varies from product to product. The consequential effects are widespread. Differential contraction can lead to warpage or cracking. Variations in density make it difficult to achieve uniformity in glazing; surface finish may suffer. Residual porosity in the fired piece may affect strength and translucency. The effects may be acceptable in products with a high degree of porosity, and unacceptable in vitreous products.

The future of dust pressing for flatware is thus still under development and review. The advantages offered include the possibility of automated rapid production, and dimensional reproducibility, which in itself offers many subsequent advantages. In the long term the possibility of using body systems not dependent on clay must be a relevant factor. On the other side has to be set the disadvantages. These include the compromise outlined above between density

Fig. 6.6 – Isostatic pressing of flatware. (Courtesy Dorst.)

Fig. 6.7 – Arrangement for dust pressing of flatware, showing dies in vertical mode. (Courtesy Netzsch.).

variation and accuracy of shape, the reduction of surface quality associated with the use of flexible dies, and the limitation of available shapes compared with plastic making.

Another problem associated with all forms of pressing is that of sticking on the face of the die. Just as there is cohesion between particle and particle in the body, so there can be adhesion between the particle and the face of the die. In general there are no shearing forces available parallel to the die face to break this adhesion. Once sticking begins, there is a progressive build-up on the die, and a corresponding deterioration of the surface finish on the pressed article. The available methods of dealing with this problem include intermittent automatic die cleaning, the use of an elastomer faced die, the introduction of a shear by slight rotation of the die, organic additives in the body, and heating the die.

Two other forms of pressing are worthy of mention, though not strictly within the scope of this chapter. One is hot pressing, often used in the production of technical ceramics, in which the powder is pressed and sintered at the same time at a high temperature. The other is plastic pressing; large flatware dishes can be formed by this method using plastic clay bodies formed under pressure on a plaster mould.

One feature of semi-dry pressing that may be noted is that the final product shows very little of the orientation of clay particles that is such an important feature of both plastic forming and casting. There is a certain degree of parallelism surviving within the original granule, but the pressing process itself introduces very little. Thus many of the troubles, such as distortion, encountered in the other processes, are much less in evidence here.

## REFERENCES

Cubbon, R. C. P. (1982) *Tr. & J. Brit. Cer. Soc.*, **81**(1), 9.
Holdridge, D. A. (1952) *Tr. Brit. Cer. Soc.*, **51**, 401.
Stuijts, A. L. and Oudemans, G. J. (1965) *Proc. Brit. Cer. Soc.*, **3**, 81.
West, H. W. H. (1955) *Tr. Brit. Cer. Soc.*, **54**, 543.

# 7

# Moulds

Before turning to the various other methods of forming used in pottery manufacture, it is necessary to consider the mould in which, or on which, the body is transformed into the required shape. Apart from such hand processes as throwing, and techniques based on dust-pressing, all other products make use of moulds in the making process. Plates and other flatware are made on moulds; hollow articles, and especially those of complicated shape, are made from slip, using the internal surface of moulds. Hollow tableware, ornamental ware, and the whole of the sanitary industry depend on this latter method. The mould is vital to the success of these operations, and any deficiencies here can result in serious process losses.

Traditionally, the most commonly used material for moulds is plaster of Paris, which has been in use for this purpose for at least three hundred years. It has a number of outstanding properties not easily matched by other materials. It is available in quantity at an economic price, and can be processed in an uncomplicated manner. It provides a smooth surface, with fine detail, which is essential for the accurate reproduction of shape. It can easily be made to simulate the most complicated profiles, and can be made into working moulds in the quantities required for the commercial repetitive manufacture of articles. The function of the mould in addition to reproducing the required profile is to extract water from the body or slip, and plaster has the ideal texture for performing this function.

## 7.1 PLASTER OF PARIS

Plaster of Paris is made from gypsum, the most abundant supply of this mineral in Britain occurring in the Newark area. The deposits may occur near the surface, in which case they can be open-mined, or they may be deeper down, in which case underground mining is necessary. The rock is carefully selected for use in the pottery industry, since much higher standards of quality are required there than are necessary for building plaster, which uses very much larger tonnages of the material. Some impurities are inevitable; some soluble and others insoluble. Insoluble impurities, such as quartz, can result in protrusions on the surface of the mould, thus spoiling the surface of the finished ware.

After some degree of crushing, the rock is heated in a container, which may be one of several kinds. The process may be discontinuous, as in a kettle or autoclave, or a continuous rotary kiln may be used. The material is heated to a temperature of about 150 °C, and this drives off some of the combined water converting the gypsum to hemihydrate. The properties of the resulting plaster depend very critically on this calcining process. Some particles may remain as gypsum, and others may be over-calcined so as to convert to the anhydrite, and either of these can influence the blending and setting behaviour. The reaction with water can also be altered by the addition of accelerators or retarders.

After calcining, the plaster is finely ground, and should preferably be aged for some time before use. The blending of freshly calcined plaster is very difficult to control.

## 7.2 MAKING THE MOULD

Mould making usually involves a series of steps which can be regarded as similar to the development and printing of photographs.

(1) The original positive embodies the shape of the article which is eventually to be produced in quantity. This model may be made, for example, by sculpting a solid block of plaster. It may equally well be made in clay, wood, metal, plastic, or any other material giving permanent form and a smooth surface (Fig. 7.1(a)).
(2) From this original model a negative cast is made in plaster, known as the block mould (Fig. 7.1(b)).
(3) From this block mould another positive is cast, in the same shape as the original model, known as the case mould (Fig. 7.1(c)).
(4) Any number of negatives, known as the working moulds, are made in plaster from the case mould. These working moulds may be used several hundred times before wear of the surface makes them unusable (Fig. 7.1(d)). The making of plaster moulds for flatware has been described in detail by Oliver and Mullington (1965).

Plaster expands slightly on setting. This is an advantage in that it improves the reproducibility of intricate detail from the model, but it can make it difficult

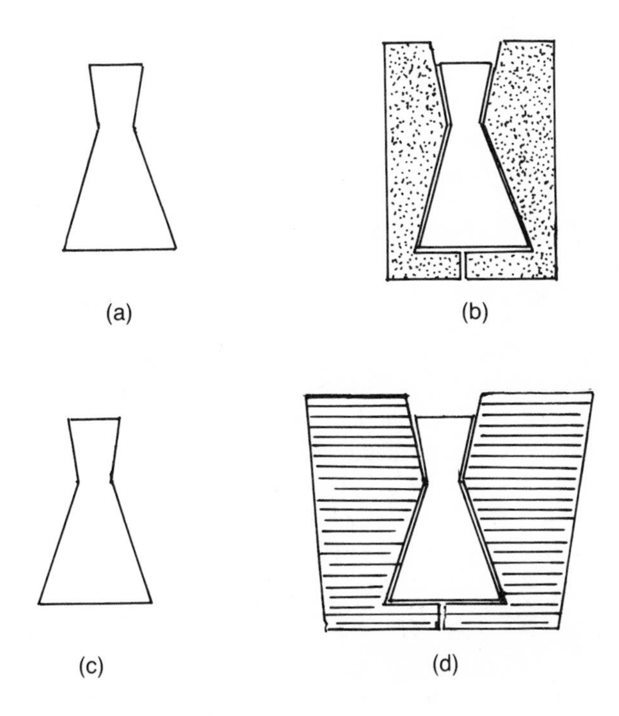


Fig. 7.1 – Stages in making plaster moulds.

to separate plaster–plaster surfaces. This can be overcome by using release agents, such as soap or other organic substances. In modern practice, the problem is alleviated by the use of resins for the block and case moulds. Resins and metals can also be used to increase the effective strength of plaster moulds.

In these mould-making processes a critically important part is the blending of the plaster and the water. Accurate proportioning of plaster and water should be aimed at, but, even so, variations in the properties of different batches of plaster are such that a standard blending programme cannot be relied on. When the plaster and water mix is stirred, hydration takes place after a few minutes, and is accompanied by a rapid increase in viscosity. The mould maker has to judge from experience the time at which to pour the mix, which must take place whilst the mix is sufficiently fluid to fill all parts of the mould. Modern blending machines incorporate automatic control. In order that allowance may be made for plaster variations, these control systems cannot operate on a fixed time basis but have to incorporate some means of measuring continuously the viscosity of the mix. Cubbon and Walker (1981) have shown how the continuous recording of the viscosity of the mix during blending can lead to a better understanding of the setting process and a basis for a blending system independent of the judgement of the operative. The British Ceramic Research Association has designed a blending unit which incorporates: automatic charging and weighing of the plaster and water; mixing under a vacuum bell; and mixing control based on viscosity change, taking account of the water demand of the plaster (Fig. 7.2). Air bubbles in the mix can result in poor mould surfaces, low mould life, and uneven finished surfaces on the ware. In order to improve mould performance in this respect, blending under vacuum can be used. Modern blending plant will thus incorporate accurate proportioning arrangements, automatic viscosity control, and high vacuum conditions. The use of such techniques can improve mould life by up to fourfold.

The drying of the working moulds can also have some influence on their behaviour in use. As the water migrates to the surface during the drying process it may carry with it soluble salts, and these can create a layer at the surface of the mould that is markedly different in texture from the bulk of the mould material. In order to avoid difficulties arising from this during manufacture,

Fig. 7.2 – Plaster blending unit. (Courtesy British Ceramic Research Association.)

it may sometimes be necessary to arrange the drying so that the migration of water is directed away from the working face of the mould (Cubbon, 1982).

## 7.3 PROPERTIES OF SET PLASTER

A number of physical properties of the set plaster determine its effectiveness in use, either for plastic making or casting, and its durability. In both cases, properties affecting the rate of water removal and mechanical strength may be expected to be of significance. An understanding of these properties may best be achieved by considering how the texture of plaster develops during the setting process.

Ignoring for the moment the influence of small amounts of impurities, and especially traces of gypsum or anhydrite, the setting of plaster may be considered in terms of the following reaction:

| | Hemihydrate | + | water | → | Set plaster |
|---|---|---|---|---|---|
| | $CaSO_4.\frac{1}{2}H_2O$ | + | $1\frac{1}{2}H_2O$ | | $CaSO_4.2H_2O$ |
| Molecular weights | 136 9 | | 27 | | 136 36 |

Fig. 7.3 – Scanning electron micrograph of a fractured plaster mould surface, showing the lath-shaped calcium sulphate dihydrate crystals. (After Cubbon. Courtesy British Ceramic Research Association.)

Water content is generally expressed in terms of the plaster:water ratio. This may vary over a range for different purposes, but a fairly typical blending would use $P/W = 100/70$, by weight. Hemihydrate has about five times the solubility of gypsum, so contact with the water produces localized regions of supersaturated solution, resulting in the formation of crystals of dihydrate. This process is repeated until rehydration is complete. Because the particles are fairly close together at these concentrations, the result is a mass of interlocking laths with a reasonable degree of strength and the required degree of openness so as to take up water. This structure is illustrated in Fig. 7.3. The true density of the hemihydrate is about $2.75 \times 10^3$ kg m$^{-3}$, and that of the dihydrate is about $2.32 \times 10^3$ kg m$^{-3}$, so a contraction on setting might be expected. In fact this does not happen. The arrangement of the crystals is such that setting is usually accompanied by a volume expansion of about 0.5 per cent. This is advantageous in that it presses the plaster into the recesses of the case mould and helps in the reproduction of the detail.

From the above equation it can be seen that 100 kg of hemihydrate need 18.6 kg of water for rehydration. Starting with $P/W = 100/70$, the set product

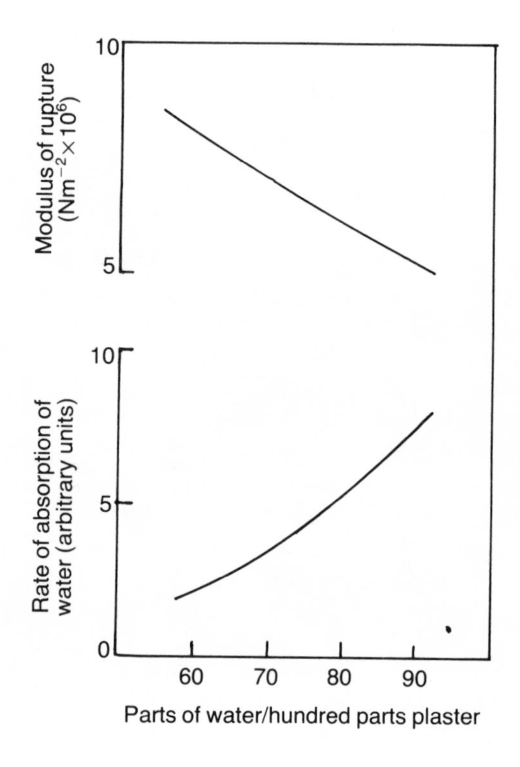


Fig. 7.4 – Properties of set plaster.

will consist of 118.6 kg of dihydrate, with a density of $2.32 \times 10^3$ kg m$^{-3}$, equal to a volume of $51.1 \times 10^{-3}$ m$^3$, together with 51.4 kg of water, or $51.4 \times 10^{-3}$ m$^3$ of pore space when the excess water has been removed. Set plaster, then, may be expected to have a volume porosity of around 50 per cent.

Porosity in itself, however, is not the whole story. The important property from a working point of view, is the suction rate, or the rate at which the mould extracts water from a casting slip or a piece of plastic body. This depends not only on the porosity, but on the size of the pores; a complex mix of surface tension, porosity and permeability. To give these parameters some quantitative expression, the following approximate values may be noted for a 100/70 blend.

Volume porosity: 50%
Internal surface area: 500 m$^2$ kg$^{-1}$
Permeability to air: $2 \times 10^{-3}$ cm$^2$.s$^{-1}$/cm water
Mean pore radius: 2 $\mu$m
Permeability to water: $10^{-4}$ cm$^2$.s$^{-1}$/cm Hg

Increased suction may be obtained only to a limited degree because of the need to maintain strength. Although some of the deterioration of moulds in service may be due to chemical attack from slips, and the finite though small solubility of set plaster in water, much is also due to mechanical factors, such as impact and abrasion. Set plaster at $P/W$ = 100/70 has a modulus of rupture of about $7 \times 10^6$ Nm$^{-2}$, and values markedly less than this would result in short mould life.

A compromise has thus to be effected between absorption, which increases with increasing water content in the mix, and strength, which decreases rapidly, as shown in Fig. 7.4. The size of the dihydrate crystals, of the order of 10–20 $\mu$m, is influenced by the stirring rate and blending time, and affects both the strength and the permeability. Long blending times, with crystallization going on as the stirring proceeds, result in a smaller crystal size and a decreased permeability.

Various faults can arise, for example, in the making of plates, and can give rise to serious manufacturing losses if the permeability of the mould is not right (Cubbon, 1982). Losses from new moulds can be caused by low permeability at the surface of the mould. Tearing or cracking of the face of the plate can result if the suction is not sufficient to prevent relative movement between the plaster and the mould. All these factors confirm the need for accurate control of stirring rate and blending time, and automatic compensation for plaster variables, in order to ensure that moulds of the correct texture are produced.

## 7.4 EFFECT OF HEAT ON PLASTER

The effect of heat on plaster is of great importance in the technology of drying, since it imposes an upper limit on practical dryer temperatures. Dinsdale (1953) has examined the dehydration process and related effects. When set plaster is heated, and after the pore water has been removed, combined water may be lost at a rate depending on the temperature and humidity of the atmosphere. Conversion from dihydrate to hemihydrate results in a serious loss of strength, and hence a much reduced mould life.

Fig. 7.5 shows the dissociation pressure of $CaSO_4.2H_2O$ at various temperatures. Also shown are the vapour pressures of the atmosphere at various values of the relative humidity. A plot of the reciprocal of the absolute temperature against the log of the pressure is approximately linear. Dehydration will proceed at any temperature if the relative humidity is low enough, that is, if the vapour pressure is less than the dissociation pressure. With relative humidity around 60 per cent, 60–70 °C would appear to be an upper limit for drying temperatures. In practice, there are mitigating circumstances. Residual pore water in the mould maintains an effective 100 per cent relative humidity whatever the value in the dryer atmosphere; short drying times help to maintain this condition. The most harmful condition is the combination of high temperature and low humidity, but since these are precisely the circumstances that offer optimum drying, it will be seen that the problem is of great practical significance.

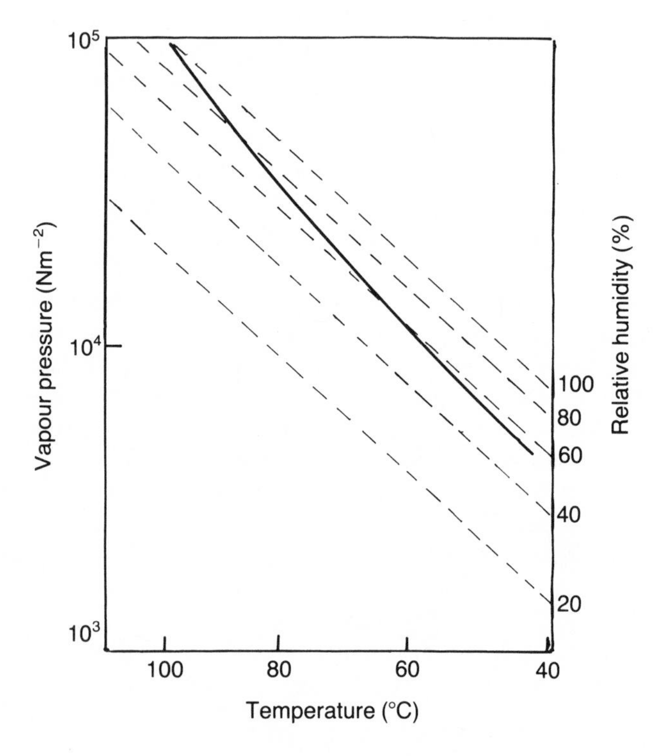


Fig. 7.5 – Dissociation pressure of set plaster in relation to vapour pressure of moist air at various relative humidities. ––– Vapour pressure of atmosphere; —— Dissociation pressure of $CaSO_4.2H_2O$.

New possibilities open up when the ware can be separated from the mould and dried on its own.

## 7.5 SUBSTITUTE MATERIALS

Many attempts have been made over the years to find substitutes for plaster as a mould material in an attempt to reduce costs and space required for mould-making and storage. In addition, reduction in mould wear would result in improved reproducibility of size and shape, together with improved surface finish.

Regeneration of old moulds by grinding and calcining has been shown to be a technical possibility under suitable controlled conditions. A range of organic resins, setting with an open pore structure have been tried. Sintered porous metal is another possibility. Good working moulds, having long life, can certainly be made from fired ceramic bodies, or from fired mixtures of clay and diatomaceous earth, for example. However, all these possible alternatives turn out to have serious disadvantages. Some of them are difficult to make in the first place, bearing in mind the need for accurate dimensions. Some have inferior water extraction properties. Some confer no cost advantage. On balance it seems that the advantages of plaster, such as cheapness, ease of working, and excellent reproduction of surface detail, are such that it will continue to be the principal mould material for pottery manufacture for the foreseeable future.

## REFERENCES

Cubbon, R. C. P. and Walker, E. G. (1981) *Tr. & J. Brit. Cer. Soc.*, **80**(3), 97.
Cubbon, R. C. P. (1982) *Tr. & J. Brit. Cer. Soc.*, **81**(6), 169.
Dinsdale, A. (1953) *Tr. Brit. Cer. Soc.*, **52**, 614.
Oliver, J. and Mullington, P. (1965) *J. Brit. Cer. Soc.*, **2**(1), 52.

# 8

# Plastic forming

The plastic working of clay is among the oldest activities of man, practised for centuries long before casting or powder pressing were thought of. In the beginning, articles were formed from natural clay, without any additions, and working was by hand only, until the advent of the wheel made other methods, notably throwing, possible. Today, plastic forming is still the most important of the making methods, accounting for most tableware products and essential also in the electrical porcelain sector.

As has been previously noted, the process depends on the unique rheological properties of clay, which dominate the character of bodies even when diluted with other materials. What makes it possible is the ability to deform easily so as to take up an imposed shape without rupture, and then to retain that shape when the deforming forces are removed. The development of a high level of strength on the application of heat is an additional important feature. In rheological terms, the significant parameters are an appreciable yield value, at least sufficient to support the article under its own weight; a low stress: strain ratio during deformation; and a long working range up to the strain value at which rupture occurs. All these characteristics are present in clay-based bodies. Their precise quantitative values can be manipulated to suit particular purposes by changing the body mix, the particle size distribution, and the moisture content. The moisture content is generally within the range 18–24 per cent, the higher end being most appropriate for throwing, and the lower end for roller-head making.

An important element in the plastic forming process is the use of plaster moulds, considered in detail in Chapter 7. This material makes for easy reproduction of desired shape, extracts some of the water from the body thus increasing the rigidity, and confers an excellent surface finish on the ware.

In this chapter we look at some of the important physical aspects of the process. These are perhaps best exemplified by the forming of dinner plates, though many of the principles are applicable to other types of ware.

## 8.1 FLATWARE

The history of flatware making, at least until the last two or three decades, has been mainly the story of the development of the jigger. Present methods have been reviewed in detail by Cubbon (1982(a)). We use the term jiggering to refer to flatware, and jolleying to refer to hollow-ware. In the jiggering process a shaped metal tool is brought into contact with the clay on a rotating plaster mould. The mould forms the profile of the front of the plate, and the tool forms the back surface. The arrangement is as shown in Fig. 8.1(a). The plaster mould is cast into a metal head, which can be inserted into a chuck attached to a rotating vertical spindle. A piece of pugged clay of the required volume is first flattened out on a batting head, and the batt is then placed on the mould. The mould is inserted into the chuck and as the head rotates the tool is brought down into such a position as to form the bottom of the plate. In doing this, it applies a moderate pressure to the body and produces a considerable shearing action. It also removes an appreciable amount of surplus body, perhaps 20 per cent, which is returned to the pug or extruder. The application of a spray of water mist during the operation helps to improve the surface finish. The mould is then removed and placed in a dryer; when some 8 or 10 per cent of the water

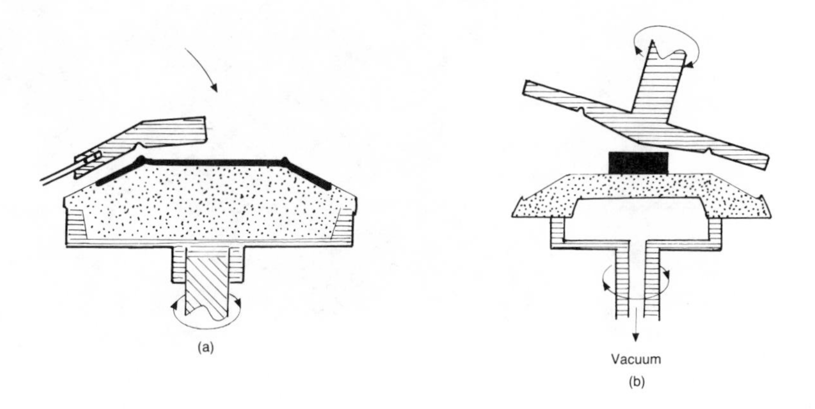


Fig. 8.1 – Arrangement of tool and mould for (a) jigger and (b) roller head.

has been removed, mould release takes place. Most of the water removed goes into the plaster mould, so it is desirable to have some further drying in order to avoid water build-up in the mould.

For many years this process was hand-operated, even when power driven heads had replaced the old treadle. The degree of pressure on the jigger tool was judged by the maker in relation to the state of his clay, and this required a considerable degree of skill and experience. During this century the semi-automatic jigger was developed, and following this the whole process became automatic. Beginning with the slice from the pug, and proceeding with the batting out and transfer to the mould; jiggering, and transfer to the dryer; and return to the mould; no operative is needed. Of course, the elimination of the skill and judgement of the operator has meant that improved consistency of materials and control of properties is called for.

These modern automatic jiggers use a mould rotation speed of between 600 and 1000 rpm, compared with 300–400 for hand jiggering. A single head will produce plates at the rate of 15-20 per minute, and a large industrial installation could have as many as eight heads.

Fig. 8.2 – Roller making machine for flatware. (Courtesy Service Engineers.)

Over the last thirty years the jigger has been gradually overtaken by the roller-head method, originally introduced for making hollow-ware, but now firmly established as the best method of making flatware. The system is illustrated in Fig. 8.1(b) and Fig. 8.2. In this case the plaster mould is hollow, so that when it is attached to the rotating head, a vacuum can be applied. This produces a suction at the surface of the mould which assists in holding the clay on the mould during forming. The tool is in the form of a driven roller, made of metal, and usually heated to prevent clay from sticking to the face. No batting-out is required. A slice of body, less in area than the mould, is cut off the pug, placed in the centre of the mould, and rolled out when the roller is brought down. As with jiggering, the roller-head process can be fully automated. The rotation speed is of the order of 500 rpm. The mould speed is slightly higher than that of the roller-head and the differential needs to be accurately maintained in order to provide the right kind of shearing action. The rate of output of flatware is of the order of 12 to 15 per minute. The method has several advantages compared with the jigger; notably, the elimination of the batting-out stage, and a superior finished product in terms of homogeneity of texture. In some installations, the drying is divided into two parts, the first stage bringing about mould release only. In the second stage the ware alone is dried, the absence of the mould enabling higher temperatures to be used. The early return of the mould to the making head reduces the number of moulds required.

We turn now to consider some of the elements of the process that can cause problems if not clearly understood and allowed for. The applied stress in the forming process is not well defined. The effective pressure is in the range 20 to 50 $Nm^{-2}$, which is slightly above the measured yield value for plastic body. The actual yield value may be quite different at these high rates of strain compared with the value obtained in experiments at very low rates of strain. For a mould rotating at 500 rpm, the linear speed at the edge of a 10 in plate is about 6 $ms^{-1}$. Against a stationary tool this means a rate of strain of many thousands per sec. The strain is also large, which is why a long working range is required. Particularly in the jiggering process, clay is transported horizontally over a considerable distance. A very important feature of both processes is the degree of orientation introduced into the texture of the finished article, (Cox and Williamson (1958)). Fig. 8.3(a) illustrates the more or less random orientation of clay particles in the slice from the pug. Some orientation will have occurred in the pug, and more in the batting out process, so the picture is not as simple as is shown. However,

Fig. 8.3 – Orientation of clay particles in plastic-made ware. (a) Before shaping; (b) after shaping.

when the body is sheared either by the jigger or the roller-head, a very considerable degree of parallelism develops. Those plate-shaped clay particles initially parallel to the mould surface remain so; many others become parallel as a result of the shear. It should also be remembered that some of the non-clay constituents are far from equidimensional, and needle shapes orientate almost as readily as plates. The resulting texture of the body in the finished plate is as shown in Fig. 8.2(b). The most important consequence of this is a marked anisotropy in the contraction. The parallel direction contains less inter-particle spaces, filled with water, per unit length than does the perpendicular direction. Consequently, the drying contraction is appreciably higher in the prependicular direction. The difference is even more marked during firing, where the contraction may be as much as 50 per cent higher in the perpendicular direction than it is in the parallel. These differences make for problems with distortion or cracking unless drying and firing are carefully controlled. The visible protrusion that often appears at the centre of a plate is a result of the lower contraction associated with less orientation in that region. There may be significant dimensional differences in large articles, such as are met with in the electrical porcelain industry. In marked contrast, this kind of orientation is almost completely absent in plates made by the powder pressing method.

## 8.2 FAULTS IN FLATWARE

Recent studies in roller-head making, reported by Cubbon (1982(b)), have thrown light on the cause of a number of common sources of loss in production, and have pointed the way to their reduction. Close technical control of the technical parameters is essential for optimum results, and among the more important are the moisture content of the body, the temperature of the roller head, and the relative velocities of the head and the mould.

### 8.2.1 Humpers and whirlers

A potential source of loss is known in the trade as the problem of 'humpers' and 'whirlers'. When the centre is deformed in the convex-upwards sense the plate is said to be a 'humper', as shown in Fig. 8.4(a); conversely, Fig. 8.4(b) shows the 'whirler', with a convex-downwards distortion, the ideal situation being a

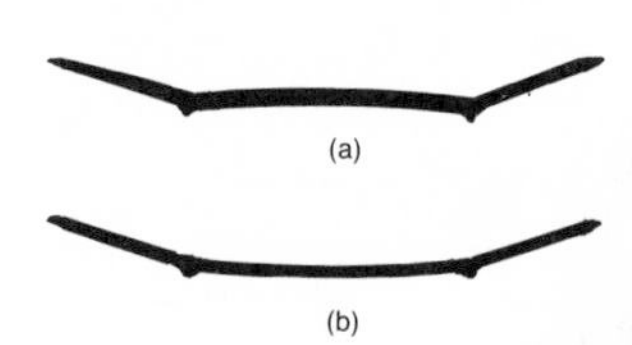


Fig. 8.4 – (a) Humpers and (b) whirlers.

completely horizontal profile. The phenomenon manifests itself after the biscuit firing, but the seeds of the trouble are sown in the making and drying. The centre of the plate always drops by as much as 1 or 2 mm, during the biscuit fire, so it is necessary to have a convex-upwards profile after drying. This is helped by making the mould with a convex surface, the so-called 'spring'. Generally speaking the loss is at a minimum when the probability of distortion is equal for the upwards and downwards direction. Any initial bias in either direction increases the fault in that direction, and diminishes it in the other, but the sum of the two increases.

The force producing the distortion is derived from the removal of water during drying and associated shrinkage. A clay plate is in the same case as any flat lamina, in that it is unstable in the presence of lateral compressive forces. In these circumstances, buckling is a well-known phenomenon. If the rim of a plate fires before the central area, the shrinkage of the rim exerts such a radial pressure, and the centre may be forced out of line, either upwards or downwards. Any kind of asymmetry or inhomogeneity exacerbates the situation. Thus the orientation of particles during making, resulting in different contractions in two perpendicular directions may have an adverse effect. Differences in moisture content from point to point produce asymmetrical forces. Slight changes in moisture content can easily alter the capacity of the body to resist distortion. The answer to the problem can only be in correct and accurate mould design, with special reference to the spring, and the minimizing of the process variations mentioned.

### 8.2.2 Stretched face

This fault takes the form of cracks in the surface of the formed plate, often spiral in form, and generally occurring near the centre. It is clear that what has occurred is that the strain in this area has exceeded the limiting strain for the body and rupture has occurred in a lateral direction. This can happen with all types of bodies, but is obviously most likely with bodies having a short working range. As the working range can be extended by increasing the moisture content, low moisture contents should be avoided.

The excessive strain results from too much movement of the body during shaping, and it is now well established that this occurs because the body slice is not held sufficiently firmly by the mould; and this in turn means that the suction at the mould face is inadequate. The effective suction at the mould face is the suction applied by the pump minus the loss through the mould. The pressure loss through the mould depends on both the thickness and the permeability of the plaster. A reduction in mould thickness over the central area, provided sufficient strength is retained, is a simple way of improving the situation but is not in itself sufficient if the mould permeability is too low. The permeability is greatly influenced by the crystal size of the plaster, which can be controlled by attention to the plaster:water ratio, and the blending time, the latter being the more important. A working compromise has to be achieved between a structure open enough to be permeable and yet strong enough to be durable. Provided that the vacuum at the pump is adequate, and the permeability of the mould is high enough, the stretched face fault is greatly minimized.

### 8.2.3 New mould loss

When a new mould is put into service, the first few plates made from it are often lost; the cause may be cracking during drying, pieces of clay plucked from the face, or air cavities. All these derive from a specific condition at the surface of the mould. It is now known that there is a layer at the surface of a new mould, caused by the migration of soluble salts during the drying of the plaster, and characterized by a very low permeability to both water and air. Due to the presence of this layer, the plate in the dryer loses much more water from the exposed back surface than it does from the front surface in contact with the mould. The resulting differential contraction is sufficient to cause cracking. There can also be sticking of the body to the mould surface, and residual air

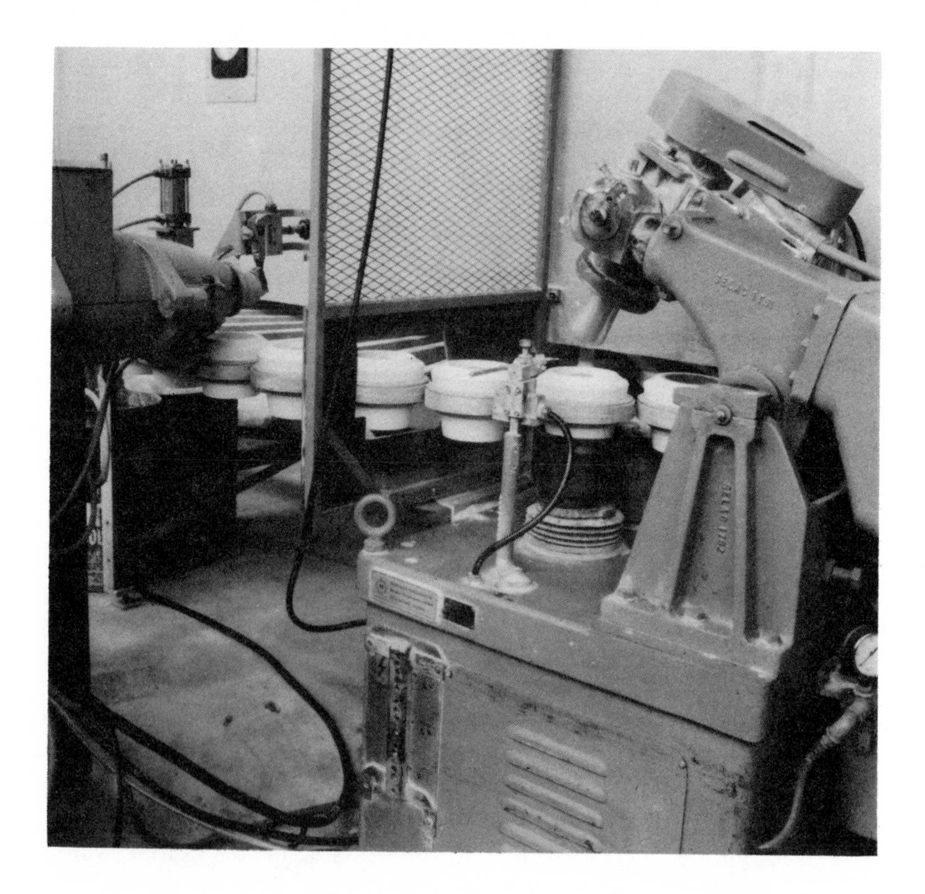

Fig. 8.5 – Automatic cup making unit. (Courtesy Service Engineers.)

blisters due to the inability of entrapped air under the body slice to escape into the mould.

After a few plates have been made the impermeable skin is worn off, and the faults disappear, but since the life of a mould is not more than 150 fills, the plates lost at the start can represent an appreciable loss of production.

The loss can be greatly reduced by ensuring that the skin is not formed, and this can be achieved by taking steps to cause all the mould drying to take place through the back surface, so that there is no migration of salts to the front face.

## 8.3 HOLLOW-WARE

The earliest uses of the roller-head process were for the production of hollow articles, such as bowls, that were not easy to jolley. The process is now commonly used for the forming of cups from plastic clay (Fig. 8.5), as is also the jolleying method. Both these can be fully automated, and continuous flow carousel systems give outputs of 20 cups per minute. The outside surface of the cup is in contact with a single-piece mould, and the tool works a slug of body from the inside outwards. In some cases, a relatively simple shape is formed, partially dried, and then turned on a lathe to give a shape too complicated to be formed in a single mould.

The problems of distortion and cracking are not so difficult as they are with plates, and the technical control requirements correspondingly less stringent.

## 8.4 PLASTIC PRESSING

There are some shapes, especially in the domestic ware sector, that are not easy to make by the jiggering process, although it is possible to jigger oval shapes. Rectangular shapes are not possible, nor are hollow articles with divisions, such as salad dishes. A method used for this kind of shape is the Ram press, so called after its inventor. The method has been described by Cubbon (1982(a)). In this the soft clay is pressed between two plaster moulds of special design. The pressure employed is of the order of 150 $Nm^{-2}$, which is several times greater than that used in jiggering, though very much less than that used in semi-dry powder pressing. For this reason, the moulds need to be stronger than those normally used for jiggering. In addition they need to be permeable, since the softer body means that there is more water to be removed. The moulds are made from a special plaster, giving high strength in spite of high permeability. Enclosed in the plaster is a series of tubes, enabling either pressure or vacuum to be applied to either top or bottom die. The moulds are held in strong metal cases, and provision is made at the rim for excess body to be extruded. Because of the complexity of the process the rate of production is relatively slow, some two or three pieces per minute. The sequence of operation can be followed by reference to Fig. 8.6, and is as follows:

(a) A slice of plastic body is placed on the bottom mould.

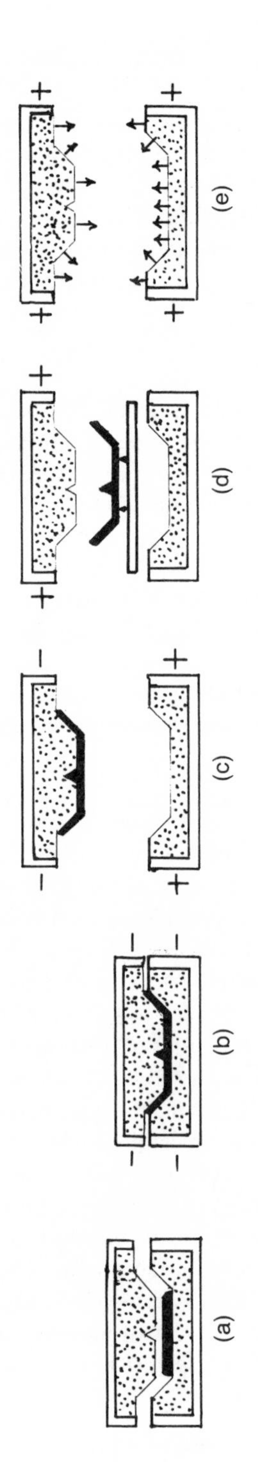


Fig. 8.6 – Stages in the Ram pressing process. + = pressure; — = vacuum.

(b) The body is pressed into the required shape. A vacuum is applied to both moulds to help to remove any entrapped air and some of the water.
(c) The top mould is raised. Pressure is applied to the bottom mould to assist mould release, and vacuum at the top to retain the pressed article on the mould.
(d) A carrier is brought into place, and the piece is forced from the mould under the action of gravity, and a positive pressure if needed.
(e) Considerable water has been absorbed by the moulds, and enough of this is now removed by the application of pressure to enable the cycle to start again.

This process has proved to be very useful for the production of large pieces or awkward shapes. Much work has been done on the development of moulds, as the imposed conditions can mean a short life. The use of fired ceramic or porous metal moulds have been tried, but plaster is still commonly used.

## 8.5 RESIDUAL STRESS

In the deformation of plastic clay there is a finite elastic component in addition to the viscous or plastic flow (Bloor, 1957). When the stress is removed, elastic recovery may take place over a period of time. The stresses imposed during a given process may give a texture pattern in the body that persists into the drying and firing stages; clay is sometimes said to have a memory. When drying takes place, the removal of the interstitial water enables particles to readjust and some elastic recovery may take place. Further movements can occur in firing. It is well known in the electrical porcelain industry that extruded and turned articles can twist after forming; sometimes in one sense during drying, and in the opposite sense during firing. This residual elastic behaviour can cause distortion, or even cracking, at any stage.

It is important to note that this memory effect is always present. Every plastic forming process produces a stress pattern in the piece, mainly due to the orientation of clay particles, and this stress pattern will persist until the elastic recovery is complete. The phenomenon is of no significance in many cases, but can be extremely troublesome with large and complex shapes.

## REFERENCES

Bloor, E. C. (1957) *Tr. Brit. Cer. Soc.*, **56**, 423.
Cox, R. W. and Williamson, W. O. (1958) *Tr. Brit. Cer. Soc.*, **57**, 85.
Cubbon, R. C. P. (1982a) *Tr. & J. Brit. Cer. Soc.*, **81**(1), 9.
Cubbon, R. C. P. (1982b) *Tr. & J. Brit. Cer. Soc.*, **81**(6), 169.

# 9

# Slip casting

Slip casting involves the use of a mould of appropriate shape into which a fluid suspension of body can be poured, and which progressively extracts some of the water until a solid layer is formed. The technique is used where it is necessary to produce complex shapes, and where plastic forming or semi-dry pressing are impossible. It is thus applicable to the production of hollow tableware, figures and ornamental ware, and is the main method of forming sanitary ware. The process in its present form is only made possible by the use of deflocculants; the effect of adding alkalis was noted nearly two hundred years ago.

Two main methods may be distinguished. They are often referred to as drain casting and solid casting, and are illustrated in principle in Fig. 9.1. In drain casting, Fig. 9.1(a), the mould is filled with slip and casting takes place on one surface only. After a suitable time, during which the required cast thickness is built up, the excess slip is poured off. The mould and cast are then partially dried, so that mould release takes place; the cast is still sufficiently plastic for any trimming, cutting, or sponging to be carried out. This process is easy to mechanize, and even automate. In the production of cups, a continuous system is used in which the moulds are carried round a circuit on a conveyor. The various automatic stages include mould filling, casting, mould emptying, drying, mould release, and further mould drying.

When varying wall thicknesses are called for, solid casting is used, as in Fig. 9.1(b). In this case the mould is filled with slip, and casting takes place on both surfaces. The removal of water generally means that the slip has to be

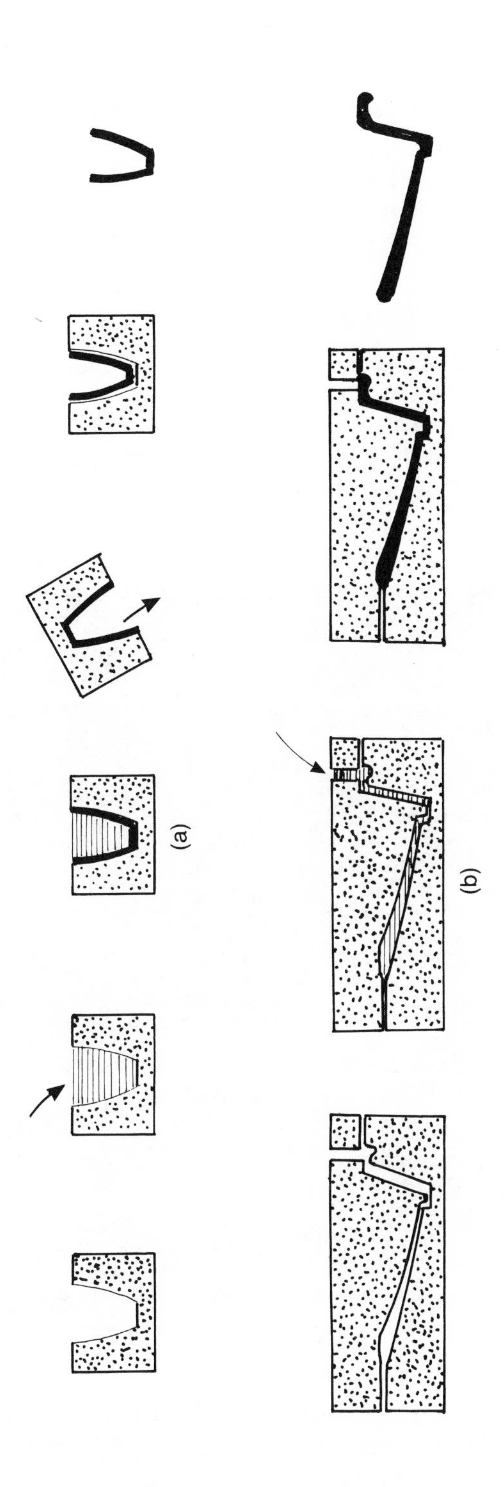


Fig. 9.1 – Stages in slip casting. (a) Drain casting; (b) solid casting.

topped up during the casting. For complex shapes the mould may have to be constructed in several sections. Most sanitary ware is made by solid casting. In some areas there have been striking increases in productivity. Mechanical conveying is well established (Fig. 9.2), but the use of so-called battery casting has resulted in another major advance. In this method the moulds for lavatory basins are arranged back-to-front on a supporting rail, similar to a filter press, and filling and emptying is accomplished without removing the moulds (Fig. 9.3).

## 9.1 PROPERTIES OF SLIPS

Reference has been made in Chapter 5 to the importance of the rheological properties of the casting slip in the casting process. In order to make the process commercially acceptable, the rate of casting has to be as high as possible. This means that high solids content slips must be used, and this is only made possible by reducing the viscosity by the use of deflocculants. As can be seen from Fig. 5.1, suitable slips can be made in a moisture content range of 27 to 35 per cent. However, casting rate is not the only criterion. The texture of the cast is impor-

Fig. 9.2 – Automatic holloware or solid casting system, incorporating Tobin process using microwaves to accelerate casting and drying. Mould filling is shown at top centre, with microwave unit on the left. (Courtesy Ceramic Microwave Products Division of Gough Engineering).

tant and this is affected by the thixotropy as well as the fluidity. The permeability of the cast influences such features as drying. Casting faults such as bad mould release, cracking, warping, surface defects, and so on, can all be caused by inappropriate slip properties, so slip control is vital. The body formulation can influence the density of the cast, which in turn influences the casting rate. Where possible, body systems should be specifically designed for casting; the use of the same body for plastic forming and for casting is less than satisfactory, although it is often practised.

One point worthy of note is that in nearly all casting operations the clay particles are deposited on the mould surface with the plates parallel to the surface. All casts are thus anisotropic to some degree, with the well known consequences in drying and firing arising from different contractions in different directions in the cast.

## 9.2 THE CASTING PROCESS

Slip casting is a complex process, and those who see it for the first time are

Fig. 9.3 – Battery casting unit for lavatory basins. (Courtesy Armitage Shanks.)

surprised at its effectiveness. Research workers have tried to understand exactly what takes place, and, in particular, there has been much debate as to whether the mechanism is purely physical, that is a filtration process, or whether there is a chemical factor as well. It is now well established that both features are involved, though filtration seems to be much more important. The relevant facts may be summarized as follows:

(1) If a casting slip is placed in a porous mould, made from either sintered metal or fired ceramic body, a cast will be formed. Clearly this is pure filtration, with little possibility of chemical action.
(2) A simple experiment shows that a cast can be formed without filtration. If a glass beaker is filled with a calcium sulphate solution, emptied and left to dry, the inside surface will be coated with a thin film of calcium sulphate. If the beaker is then filled with casting slip, a thin but rigid cast will form. Since no water has been removed, the moisture content of the cast is the same as that of the slip. Since the cast is rigid, some degree of flocculation must have occurred.
(3) The migration of mould material into the cast can be demonstrated experimentally. If radioactive calcium is used in the plaster, it can be traced in the cast, and in fact it travels through about half the thickness of the cast. The reason why it does not travel through the whole thickness may be that there is a counter-flow of water into the mould as casting proceeds.

It is thus clear that in the normal casting process we have both physical and chemical elements at work. Ion migration may be part of the reason why the rheological properties of the slip are important. However, for practical purposes the filtration process is much the more important in determining the casting rate, which is of very great industrial significance. Accordingly, we proceed now to an examination of the physical dewatering process (Walker and Dinsdale, 1959).

The process depends on a number of factors, most of which can be studied experimentally. They include:

(1) The solids concentration in the slip,
(2) The permeability of the cast,
(3) The pore size distribution in the mould, which determines the suction and the permeability.

The casting rate is essentially the balance between the suction of the mould and permeability of the cast and the mould. A wide range of experimental techniques have been used in these studies, but much can be achieved by a simple system such as a glass tube and a plaster plug. Measurement of the amount of water passing and the rate of build-up of the cast, together with information on the mould texture, afford a basis for understanding the process.

## 9.3 GENERAL THEORY

We consider a system such as that shown in Fig. 9.4. In the most general case water moves under the influence of capillary suction in the mould, together with an applied external pressure (or vacuum). The flow is resisted by the plaster plug and the cast; both these resistances increase with time as the mould fills with water and cast builds up. No account is here taken of concentration gradients as these do not invalidate the general conclusions.

Let $Q$ = quantity of water crossing unit area of cross-section perpendicular to the direction of flow; $t$ = time; $\eta$ = viscosity of liquid; $k$ = permeability, depending only on the porous medium; $K$ = effective permeability, depending on the porous medium and the liquid, with $k = \eta K$; $S$ = pressure difference due to surface tension at the advancing water front; $P$ = any applied pressure or vacuum; $\mathrm{d}P/\mathrm{d}l$ = pressure gradient in direction of flow; $\sigma$ = volume concentration of solid in slip; $\rho$ = porosity of cast or mould; $l_c$ = thickness of cast; $l_m$ = depth of penetration of water into mould; and suffixes c and m refer to cast and mould. Then, considering the growth of the cast

$$Q = l_c \left[ \frac{(1-\rho_c)}{\sigma} - 1 \right] = l_m \rho_m$$

From Darcy's law

$$\frac{\mathrm{d}Q}{\mathrm{d}t} = K_c \left( \frac{\mathrm{d}P}{\mathrm{d}l} \right)_c = K_m \left( \frac{\mathrm{d}P}{\mathrm{d}l} \right)_m$$

For the pressures

$$S + P = l_c \left( \frac{\mathrm{d}P}{\mathrm{d}l} \right)_c + l_m \left( \frac{\mathrm{d}P}{\mathrm{d}l} \right)_m$$

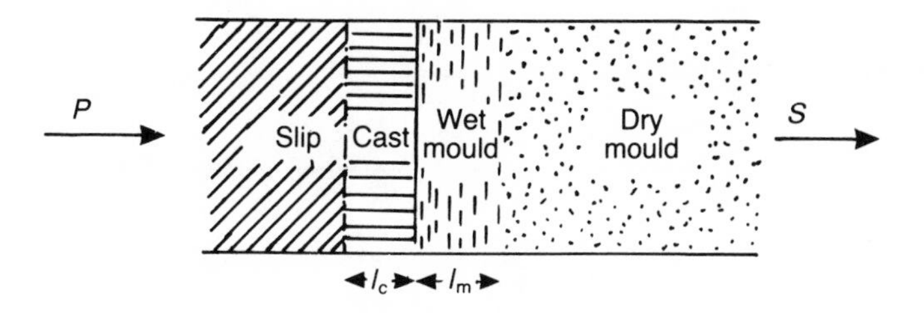


Fig. 9.4 – The general system slip-cast-mould.

From these we obtain the equation of flow

$$Q^2 = \frac{2(S+P)}{1/K_c\ [(1-\rho_c)/\sigma - 1] + 1/K_m\rho_m} \times t \qquad (9.1)$$

### 9.3.1 The system water: mould

We can first test the parabolic $Q-t$ relationship in the simplified system water: mould. For this, equation (9.1) reduces to $Q^2 = 2SK_m\rho_m \times t$. For any given mould, experiments show a very good straight-line relationship between $Q$ and $\sqrt{t}$.

It is, however, pertinent to enquire what effect the texture of the mould has in this context. Fig. 9.5(a) shows that the rate of water absorption increases with the water content of the plaster mix. As the plaster mix is made less dense, the porosity increases and so also does the permeability, but the suction due to surface tension decreases. Porosity and permeability are not simply related; increased porosity generally means increased pore size. A useful parameter is the specific permeability, which is the permeability divided by the porosity. For a plaster mould, this varies with the plaster:water ratio in the manner shown in Fig. 9.5(b). The more dense moulds have a lower specific permeability, indicating a smaller pore radius.

We thus have the situation that as moulds are made more dense, the pores become smaller and the resistance to flow increases. On the other hand, the suction in a pore increases as the radius is reduced. On balance the gain in suction is outweighed by the increase of resistance to flow, and the flow decreases.

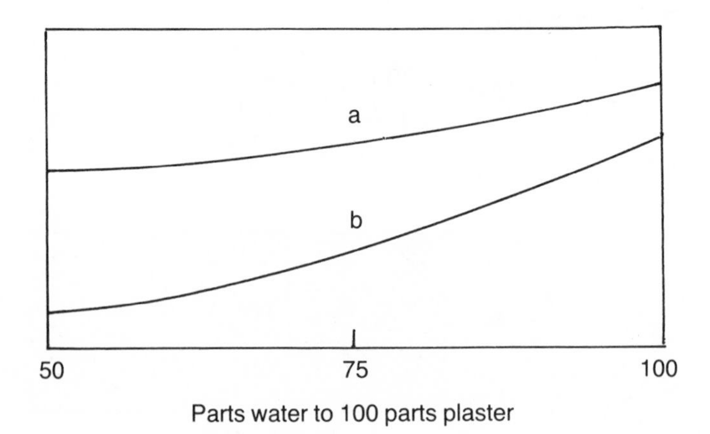


Fig. 9.5 – Water transport in plaster mould. a, Rate of water absorption; b, specific permeability.

A useful simple analogy will serve to show why. Pore structures in moulds are not simple, but they can be represented to some degree by a collection of cylindrical tubes. If we imagine water in a bundle of glass tubes, of radius $r$, in parallel, we can note the following characteristics:

Viscous flow through one tube $\propto r^4$.
Suction in one tube $\propto 1/r$.
Hence flow through one tube $\propto r^3$
Number of tubes per unit area $\propto 1/r^2$.
Hence total flow $\propto r$.

Dense moulds are desirable from many points of view, including the need for strength, but the rate of water flow through them in this simple system is diminished.

Incidentally, there has been much speculation about the hydrostatic suction in a mould. It is impossible to calculate this, since the pores are tortuous and there is a wide range of pore radii. However, if we take the surface tension of water as 73 dynes. $cm^{-1}$, it is easy to calculate the meniscus pressure for various pore sizes, thus

| | | | |
|---|---|---|---|
| at $r =$ | 0.1 $\mu$m, | pressure = | 1.5 MN $m^{-2}$ |
| | 1 $\mu$m | | 0.15 |
| | 10 $\mu$m | | 0.015 |

Pore sizes in plaster moulds frequently range from 0.1 to 10 $\mu$m, so it may be assumed that the hydraulic pressure is sometimes less than atmospheric (0.1 MN $m^{-2}$) and sometimes greater. Dal and Berden (1968) have analysed the capillary action of plaster moulds in considerable detail, including both theoretical and experimental aspects.

### 9.3.2 The system slip : cast : mould

Having seen that the simple mould : water system obeys the $\sqrt{t}$ law, and that the texture of the mould influences the flow rate in a predictable manner, we turn now to examine the much more practically significant system including the slip and the cast. In this case the cast builds up with time and offers an increasing resistance to flow. Equation (9.1) would still predict a $\sqrt{t}$ relationship, and this turns out to be so in practice. There is an abundance of experimental evidence to show that both the rate of passage of water and the build-up of cast thickness are well correlated with the square root of the time.

However, when we come to consider the effect of mould texture, the matter is somewhat more complicated. When the plaster : water ratio is varied by adding more water, the casting rate first increases, passes through a maximum, and then decreases. The situation is that there are two resistances in series opposing the flow, those of the mould and the cast. These resistances can be measured in the tube and plug experiment by the following two procedures:

(1) Use water only; saturate the mould, and apply a vacuum. Surface tension forces are now absent, and the mould resistance is constant. $Q$ is proportional to $t$, and the reciprocal of the slope of this line is a measure of the resistance of the mould.

(2) Use a slip; saturate the mould, and apply a vacuum. Combining the results of this experiment with those from (1), the resistance of the cast can be determined.

The results for a pure kaolinite slip, giving a relatively open cast are shown in Fig. 9.6. As the plaster:water ratio is increased the resistance of the mould increases, but the resistance of the cast decreases. The high rate of water flow with the open moulds results in a densely packed cast, and vice versa. The sum of the two resistances passes through a minimum. The reciprocal of this sum, which represents the casting rate, is shown by the dotted line. On the dense mould side of the maximum the process is dominated by the resistance of the mould; on the other side the resistance of the cast is the dominant feature.

In considering the practical implications of this, it is important to note that the composition of the slip plays an important part. Earthenware casting slips can give densely packed casts. The permeability of these casts can be a hundred, or even a thousand, times lower than that of the mould. We can see the effect of this by taking the data in Fig. 9.6, and recalculating the casting rate for a consant

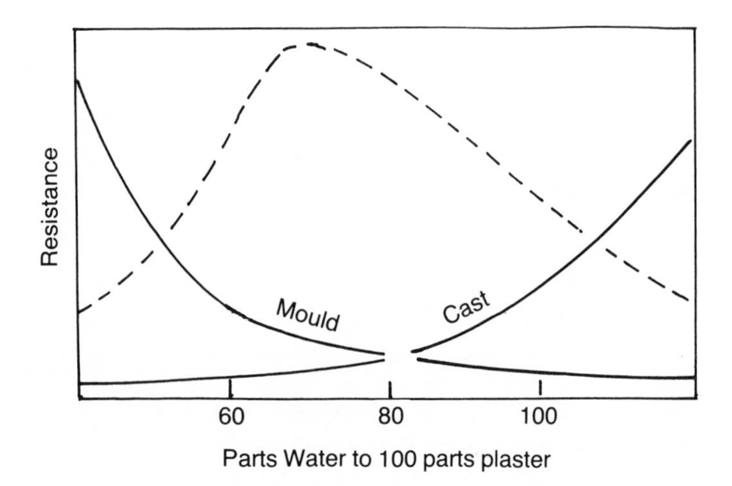


Fig. 9.6 – Mould and cast resistance for a kaolinite slip.
– – – – – – Reciprocal resistance.

mould resistance, but multiplying the cast resistance by the factors indicated, the result being shown in Fig. 9.7. Again, the casting rate passes through a maximum, but the plaster: water ratio at which the maximum occurs varies with the cast resistance. We can see a practical illustration of this in the measured casting rates of two different slips shown in Fig. 9.8. The absolute values of the casting rate have been adjusted to make clear the shape of the curves. For the pure china clay slip, curve A, the maximum occurs at a plaster: water ratio of about 100/75. For a high density sanitary slip, curve B, the maximum occurs at a plaster: water ratio of about 100/45. Since it is not easy to make moulds with a plaster: water ratio higher than 100/60, in practice casting with this type of slip will always be represented by the right-hand side of the curve. Here the process is dominated by the cast resistance and the casting rate increases with mould density. The fact that dense moulds may give higher casting rates has to be set against increased weight and cost, so that a compromise is usually accepted. This

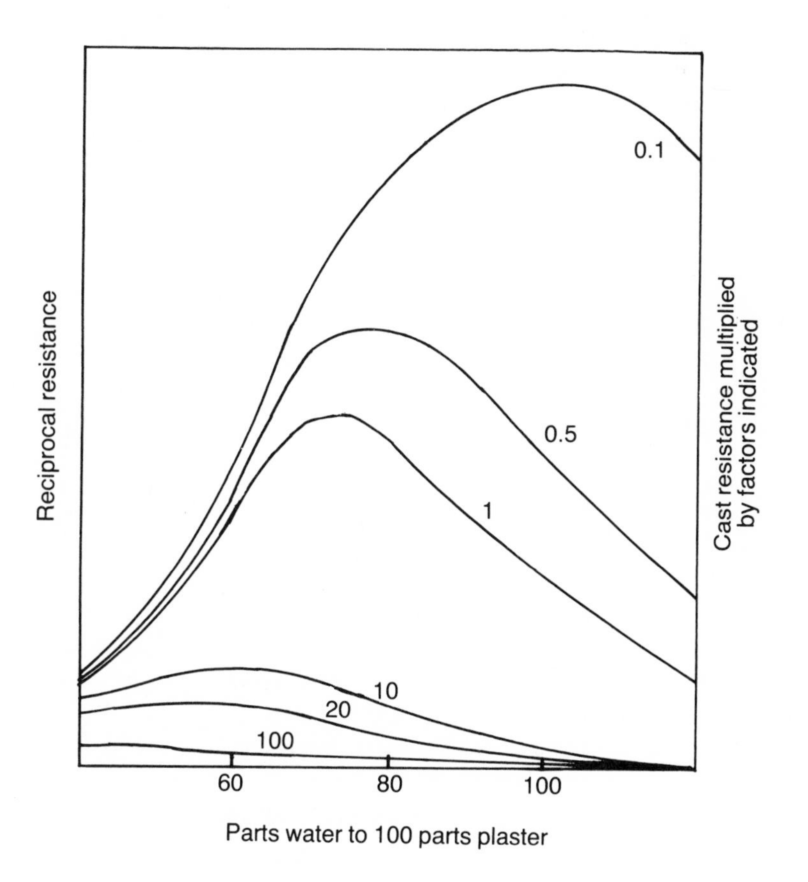


Fig. 9.7 – Effect of cast density on casting rate.

will be true for most industrial casting of clay-based bodies, but it should be remembered that the casting process is increasingly being used in special types of ceramics, where very different considerations may apply. The whole argument needs to be applied in any particular case.

In addition to the effect of the slip characteristics, the type of plaster used can have a considerable influence. In particular, additives can change the permeability, but the maximum in the casting rate still occurs.

## 9.4 TEMPERATURE

The variation of the properties of water with temperature might be expected to affect the casting rate; in particular the driving force of surface tension decreases and the viscous resistance reduces as the temperature increases. The values are as follows:

| *Temperature* (°C) | *Surface tension* | *Viscosity* |
|---|---|---|
| 15 | $73.5 \times 10^{-3}$ $Nm^{-1}$ | $1.1369 \times 10^{-3}$ $kg\,m^{-1}\,s^{-1}$ |
| 20 | 72.75 | 1.0019 |
| 30 | 71.2 | 0.7982 |
| 40 | 69.6 | 0.6540 |
| 50 | 67.9 | 0.5477 |

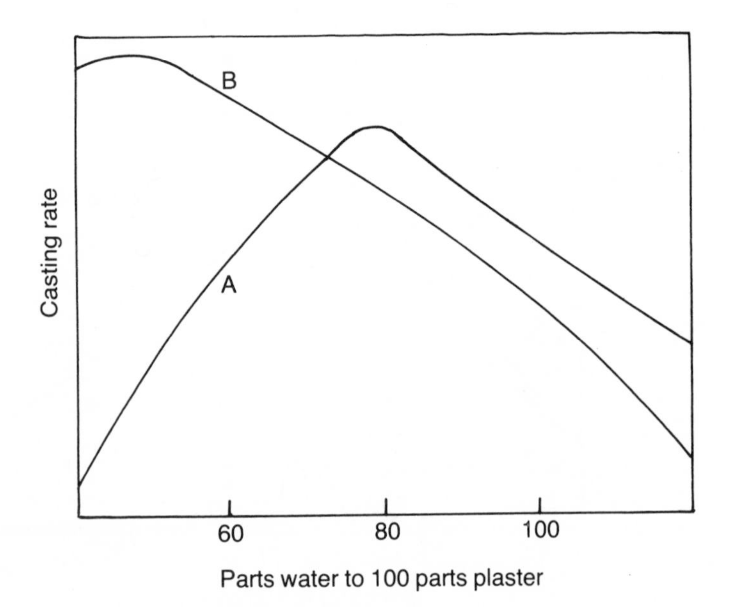


Fig. 9.8 – Measured casting rates for different slips. A, China clay slip; B, sanitary casting slip.

It will be noted that the viscosity varies much more rapidly than the surface tension; from 15 to 40 °C the surface tension decreases by 5.3 per cent while the viscosity decreases by 42 per cent. The net effect is that the casting rate increases as the temperature rises, and practical results show a good correlation between casting rate and viscosity.

In practice, there is an upper limit around 40 °C to the slip temperature before gelling occurs. There are also problems with handling, and in maintaining the required temperature in storage arks. Nevertheless, there are circumstances in which the gain in casting rate is worthwhile, and there are large scale industrial casting installations working at temperatures above normal. A gain of the order of 2 to 4 per cent in the casting rate can be achieved for each 1 °C increase in slip temperature.

## 9.5 PRESSURE

We have already noted that the hydraulic pressure operating in normal casting is of the order of one atmosphere. The possibility of increasing casting rate by the application of a positive pressure to the slip has been increasingly investigated in recent years. Moderate increases in pressure can be used with normal moulds, but beyond this it is necessary to use strong mould supports, and mould materials other than plaster can be used. For very high pressures, the strength requirements make the economics unattractive, but pressures of around 20 atm are practicable and give a rewarding reduction in casting time. Machines using porous plastic moulds are now in industrial use, reducing the casting time for lavatory basins by about 75 per cent. The possibility of complete automation makes the pressure casting system an interesting possibility for the future.

## 9.6 ELECTROPHORESIS

Electrophoresis, derived from the Greek, means carrying of electrical charges. It is well known that charged particles in a suspension can be caused to move under the influence of a potential gradient, and it might be expected that clay particles would behave in this way. In fact, if electrodes are placed in a casting slip, a cast rapidly builds up on the anode when a potential is applied. In theory, then, if the anode is shaped as a mould, it should be possible to cast in this way.

Much development work has been done on this process in recent years. It is established that suitable casts can be made, and that rapid casting rates can be achieved. However, the difficulties are formidable, and it is not likely that the process will ever be used on a large scale.

With water-based suspensions, it is difficult to avoid hydrolysis and the build-up of bubbles, so that non-aqueous solutions have to be used. The achievement of uniform thickness in the cast layer when complex shapes are used is also difficult. For these reasons the process is likely to be restricted to the fabrication of special products, such as, for example, the deposition of ceramic coatings on metal substrates. One point to note is that the slip is depleted of solids, which is not the case with traditional drain casting.

## REFERENCES

Dal, P. H. and Berden, W. J. H. (1968) *Sci. of Ceramics,* **IV**, 113.

Walker, E. G. and Dinsdale, A. (1959) *The A. T. Green Book,* 142. (British Ceramic Research Association).

# 10

# Drying

This chapter is concerned with drying, and not with dryers. Nevertheless, a brief word about dryers may be useful by way of introduction.

After the forming process, the article may have a moisture content, on the wet basis, as low as 5 per cent for pressed ware and as high as 30 per cent for cast ware. The need to remove some of the water is different for different types of ware according to the fettling processes involved. For some of these processes the body needs to be fairly plastic; sponging the edges of cups, fitting handles, and cutting apertures in sanitary ware, are examples. In these cases, more drying is needed after the fettling process. In other cases, such as the towing of plates, the ware needs to be nearly dry. Drying is necessary in order to achieve an adequate unfired strength; as noted in Chapter 6, Fig. 6.3, the strength increases rapidly with the removal of the last 2 or 3 per cent of moisture. Drying cannot normally be carried out in the kiln, as the rate of rise of temperature, dictated by the firing requirements, is too drastic. There are several major constraints on drying rate. If the article is still in contact with a plaster mould, there is an upper limit to the temperature to which the mould can be taken; deterioration of plaster can set in at 50 °C, although in the presence of moisture temperatures up to 70 or 80 °C can be used. If the heating rate is too high, pressure builds up inside the ware before the water can get out, and rupture can occur. In any case, rapid rates of heating mean increased temperature and moisture gradients, with the attendant risk of warpage or cracking.

On the other hand, there is need for speed, and this is the reason for the use of dryers. Ware will, of course, dry when left exposed to a normal atmosphere at room temperature, but it takes an unacceptably long time. The first development, then, was to place it in some kind of heated stove to speed up the drying. Continuous systems were then developed in which the ware was progressively moved through a heated chamber, either on a turntable, conveyor belt, vertical mangle, or some such means of transport. Various methods of supplying heat can be adopted, making use of conduction, radiation, and convection, and even high frequency radiation. High velocity air supplies greatly speed up the process, being particularly effective in the jet drying of cups, in which a jet of air is injected into the inside of the cup.

In these days the tendency is more and more to integrate the drying arrangement with the other processes, interposing it between making and fettling in such a manner that flow production can be achieved.

The heat requirement of the drying process is of some interest. It is an unfortunate, but inescapable, feature of drying that it usually involves the evaporation of the water, when all that is basically needed is to remove it in liquid form. The evaporation process demands a large amount of latent heat to be supplied which is not recoverable, so the thermal efficiency of dryers is not very high. Attention has to be paid to insulation and air recirculation in order to keep the energy requirements within reason. Even so, something of the order of 50 kg of steam, or its equivalent, is needed in practice to evaporate 1 kg of water in a typical dryer.

The basic requirement may be calculated as follows. Suppose we have a dinner plate weighing 0.5 kg, and we need to reduce the moisture content from 25 per cent to 5 per cent. We thus have 0.375 kg of body, and 0.125 kg of water, reducing to 0.025 kg of water at the end of drying. The specific heat of water is 4.18 kJ $kg^{-1}$ $°C^{-1}$; that of body is approximately 0.8 kJ $kg^{-1}$ $°C^{-1}$; and the latent heat of evaportion of water is 2.27 MJ $kg^{-1}$. Thus the heat to be supplied in our example is

To heat the water from 20 to 80 °C = 31 kJ
To heat the body from 20 to 80 °C = 94 kJ
To evaporate the water = 227 kJ

so that the total heat requirement for this plate is 352 kJ, the greater part of which is required for evaporation.

## 10.1 EVAPORATION OF WATER

In order to understand the drying process it is first necessary to consider the simple case of evaporation from a free water surface. Suppose we have a volume of water at a temperature $t_1$, in contact with still air at a temperature $t_2$, as shown in Fig. 10.1(a). The vapour pressure, usually denoted by the letter $e$, is to be associated with the particular temperature. The saturated vapour pressure, in equilibrium with water, increases with the temperature in the manner shown in Fig. 10.1(b), reaching atmospheric pressure at the boiling point. If the air is not saturated, the vapour pressure $e_2$ may be less than $e_1$, although $t_2$ may be greater than $t_1$. In these circumstances evaporation takes place, and the rate of evaporation is proportional to $(e_1 - e_2)$. At the same time, heat will be transferred from the air to the water, providing the latent heat of evaporation. This instantaneous state of affairs will not persist in the absence of any external heat supply. As time passes the temperature of the air will decrease and evaporation will cease. Equally, if heat is supplied only to the water, and there are no air changes, the relative humidity of the air will rise until $e_2 = e_1$ and again evaporation will cease.

Thus, in order to ensure that evaporation is to continue it is necessary to supply heat, and to replace the air so as to remove the vapour, and both these can be accomplished by introducing a flow of warm air across the surface of the water, as shown in Fig. 10.1(c). In this case a dynamic equilibrium is established, the air continuously supplying enough heat to evaporate the water, and the air change sustaining the vapour pressure gradient.

However, it is found experimentally that the evaporation rate is slowed down by the presence of a stagnant layer of air, which may be several mm thick, immediately adjacent to the water surface. Since the conductivity of air is very low, the heat flow to the water is reduced by the presence of this layer. If the air velocity is increased until turbulence sets in, the stagnant layer is broken up and removed. It is then found that the rate of evaporation increases with the air velocity, roughly in relation to $V^{0.8}$.

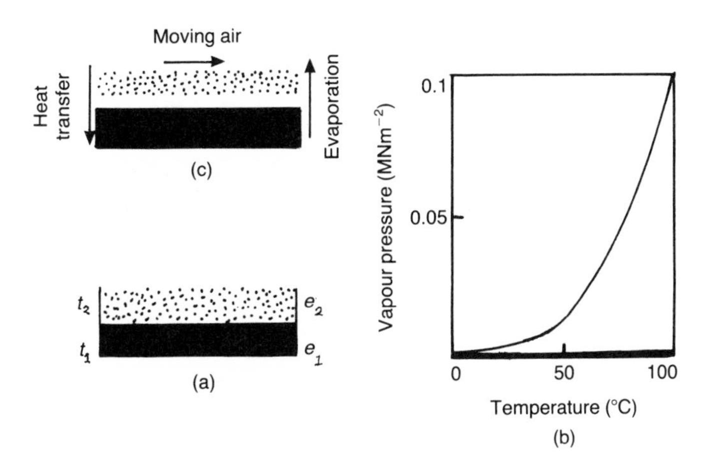


Fig. 10.1 – Evaporation from a free water surface.

So the basic requirements for a good drying rate are

(1) A high temperature of the air to provide the latent heat.
(2) A low relative humidity of the air to maintain the pressure differential and a sufficiently high velocity of the air to give turbulence and effective vapour removal.

## 10.2 EVAPORATION FROM CERAMIC BODY

A ceramic body differs from a volume of water in several critical respects, so it is necessary to look at the drying process in the practical case in some detail.

If a known weight of body, with a known moisture content, is subjected

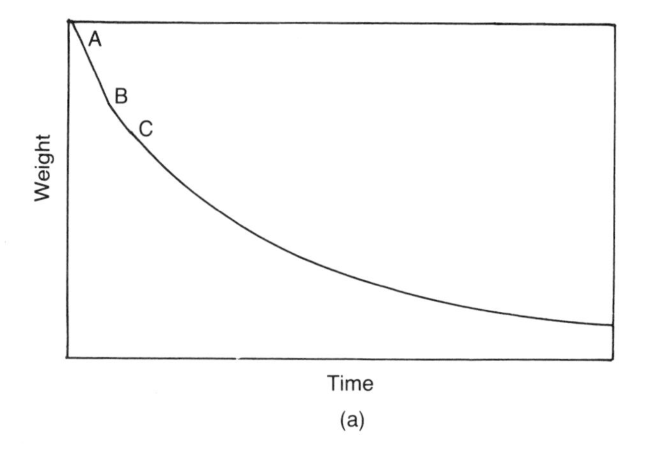


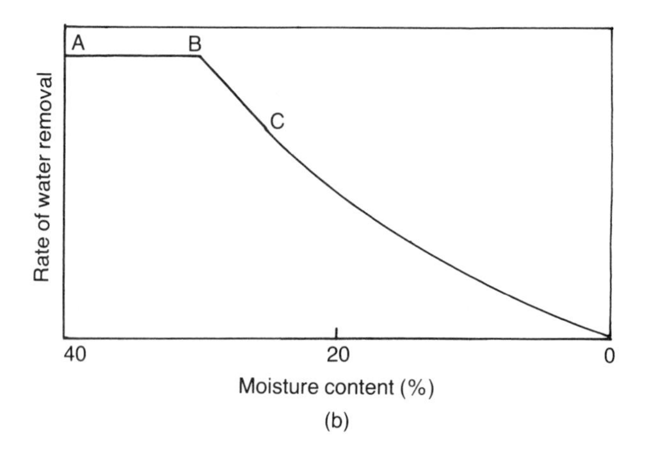


Fig. 10.2 – Drying of a ceramic body.

to drying in a warm air-stream, and if the weight is recorded at intervals of time, a graph of weight against time can be plotted. A typical curve for a white-ware body is shown in Fig. 10.2(a), after Moore (1961). If the weight of the dry body is subtracted from the ordinates, the rate of loss of water per unit time can be calculated, and can be plotted against the moisture content, as shown in Fig. 10.2(b). It will be noted that this curve has three distinct portions. From A to B, the drying rate is constant. From B to C the rate of drying falls off in a linear manner. From C onwards the rate of drying continues to fall off less and less rapidly as the dry condition is approached. The actual shape of this curve varies considerably with the type of body, being very sensitive to the pore size distribution. In some cases the point C is not discernible, but there is always a constant section and a falling rate section. The point where the one merges into the other is of considerable significance, as will become apparent when we consider the associated size changes.

There has been much debate, and experimentation, as to the physical basis of the changes in measured drying rate. Some aspects are still obscure, but there is general agreement about the main features. Possible mechanisms have been reviewed by Moore (1961). In principle, what is involved is an understanding of the balance between the rate of evaporation from the surface, and the transport of moisture through the body keeping the surface continuously supplied. In the constant rate period, AB, the pores are completely filled with water, and the surface of the body behaves in a manner similar to a free water surface. As seen in Section 10.1, the rate of evaporation depends on the temperature and pressure differentials, and on the velocity of the air-stream. As the water is evaporated from the surface, there is continuous flow of liquid water flowing through the body pores to replace it.

In the falling rate period, BC, the pores are no longer completely filled with water, and the surface is not a completely continuous water surface. There are areas which are beginning to dry, so that the rate of evaporation falls off. Water is still flowing through the body to the surface in the liquid form. Beyond the point C, in many parts of the body there is no longer a continuous liquid phase, and transport within the body is by evaporation inside the pores and diffusion through them. This process is much slower; the drying rate slows down; and the surface of the body becomes dry.

The mechanism of water transport in this region is complicated, and beyond the scope of this book. It may be noted, however, that since surface tension is involved, and since the negative hydrostatic pressure in a capillary is dependent on the radius, it is clear why the pore size distribution is an important factor. The special characteristic of clay in respect of surface adsorption of water, means that the drying of clay-containing bodies is markedly different from that obtaining in the case of other, so-called inert, granular structures.

## 10.3 MOISTURE GRADIENTS

Moisture gradients within the body are undesirable from some points of view, but inevitable, since if they were absent the water would not flow out of the system. If they are too great, they can cause severe problems with warping and cracking. In practice, they can amount to several per cent moisture content per centimetre of distance. They are greater the greater the rate of drying, so that they represent an upper limit to high drying speeds.

The precise shape of the moisture distribution curves is a matter of some complexity, involving assumptions about the mechanism of water transfer, and the special condition of water in contact with clay particles. However, a general pattern is well established and the general characteristics are represented in Fig. 10.3. In this diagram the horizontal axis represents the distance into a body being dried from one surface, and the vertical axis represents the moisture content at a given point at various stages in the drying. The curves show the moisture content distribution at arbitrary intervals of time, working from the top of the diagram downwards as the drying proceeds. When drying commences, the distribution has the form of a horizontal line, but as water begins to evaporate from the surface a gradient is established, the moisture content being higher at the surface than at the centre. The gradients become steeper until the end of the constant rate period is approached, when they begin to decrease. Over this period, all the water is transported through the body in the liquid phase. Theoretical treatment of this process generally begins by assuming that it is analogous to the conduction of heat through a solid, with an aqueous conductivity taking the place of the thermal conductivity. However, the problem is complicated by the fact that, whereas the thermal conductivity does not vary much with tempera-

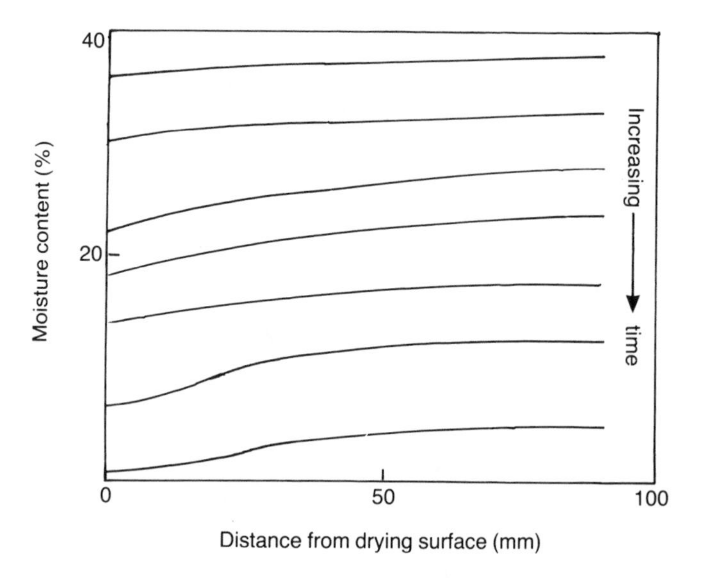


Fig. 10.3 – Moisture gradients during drying.

ture, the aqueous conductivity varies considerably with moisture content, thus making mathematical solutions difficult. In spite of this the general trend is well confirmed by experiment, and the distribution curves are roughly parabolic in shape, as the simple theory would predict.

After the constant rate period, and especially in the second falling rate period, the mechanism of transport is very different, the water now moving in the vapour phase. As the end of the drying is reached, steeper gradients occur just inside the surface as the surface moisture content drops rapidly. Since these gradients occur after contraction has ceased they are of no great practical importance.

The above considerations apply in the main to drying by warm air; if radiant heat is used the conditions at and near the surface can be markedly different. In the extreme case, if the surface is dried too quickly to the leather-hard condition, it can be difficult for the internal water to escape. Thus the rate of drying can be slowed down by the application of excessive heat at the surface.

It is of interest to note that once moisture gradients have been set up in a body, they can take a very long time to equalize out after drying has ceased.

## 10.4 TEMPERATURE GRADIENTS

Temperature gradients are not as important as moisture gradients, but they do exist when an article is heated, and they do have some effect on the flow of water. In the lower moisture content regions, an evaporation-condensation mechanism gives rise to a transport of water down the temperature gradient from the hot face to the cold, in opposition to the movement down the moisture gradient towards the hot face. This phenomenon can give rise to sharp stepwise changes in moisture distribution.

In some drying systems, attempts are made to avoid the coexistence of temperature and moisture gradients, by heating the piece up under conditions of high humidity, and subsequently reducing the humidity and keeping the temperature constant. This is the so-called humidity drying.

## 10.5 EFFECT OF THICKNESS

Many ceramic articles are complicated in shape. Electrical porcelain components and sanitary ware, in particular, have abrupt changes of contour and thickness. In these cases, it is virtually impossible to avoid changes in moisture content from point to point, and drying usually has to be slowed down so as to minimize the problem.

## 10.6 SIZE CHANGES

Ceramic bodies, when dried from their condition after casting or plastic forming, shrink on drying. This shrinkage is important in itself since it determines the relationship between the mould size and the finished size, and has to be allowed for in design stages. It also has an importance in that, in the presence of moisture gradients, it is associated with the development of stresses which may be large enough to cause distortion or even cracking.

The characteristic shrinkage curve is illustrated in Fig. 10.4, which shows the change in size of a test piece of earthenware as it dries down from the making condition to zero moisture content. The horizontal axis shows the percentage moisture content by weight, on the wet basis. The vertical axis shows the relative length of the test piece, taking the dry length as a base.

At this point, we need to clarify the relationship between volume contraction and linear contraction; the former of significance in understanding the process, but the latter perhaps of more practical importance. If we consider a cube of volume $V$, with a length of side $L$, and if the length of the side is changed by a fractional amount $\Delta L$, giving rise to a fractional volume change $\Delta V$, then it can be shown that

$$\Delta V = (1 + \Delta L)^3 - 1 \qquad \text{and} \qquad \Delta L = (\Delta V + 1)^{\frac{1}{3}} - 1$$

For small values it is usually sufficiently accurate to take $\Delta L = \Delta V/3$; the error involved may be seen in the table below.

| $\Delta V$(%) | $\Delta V/3$(%) | *True* $\Delta L$ |
|---|---|---|
| 1 | 0.33 | 0.33 |
| 2 | 0.66 | 0.66 |
| 5 | 1.66 | 1.64 |
| 10 | 3.33 | 3.23 |
| 20 | 6.66 | 6.27 |

Below the graph in Fig. 10.4 are shown the relative volumes of solid body, pores, and water at each of a number of selected moisture content values. This particular body has an unfired porosity of about 27 per cent. At the beginning of the drying process, during the constant rate period, all the pores are filled and there is excess water. As this water is removed, the body particles move closer together, and the decrease in volume of the piece is equal to the volume of water removed. At about 13 per cent moisture content there is an abrupt change in shape of the curve; this point of inflexion is known as the critical moisture content, and is a characteristic of the particular body texture. Further drying beyond this point takes place with virtually no contraction. The critical moisture content is not always a sharply defined point; with some bodies, depending on the texture, the curve undergoes a more gradual change, but the two main features are always present, namely, a region of shrinkage and a region of constant size.

It now becomes clear why moisture content differences in an article are so important. If two points have different moisture contents to the extent that one is above and the other below the critical value, it follows that on further drying one undergoes a size change and the other does not. Stresses are thus set up in the body. Below the critical moisture content drying becomes relatively easy. Powder pressed articles are usually well below this value when fabricated.

Whilst the above analysis of the contraction mechanism is broadly accepted, there is an interesting departure in detail. Below the critical moisture content value, there is often a further small but finite contraction followed by a slight expansion to the final size. The explanation for this feature is probably not simple, but is almost certainly linked with the changing significance of surface tension forces. When there is an excess of water, with all the pores full, the removal of water results in a shrinkage directly related to the volume of water removed, as has been noted. As soon as the critical moisture value is passed, some of the pores are partially empty, and surface tension forces begin to

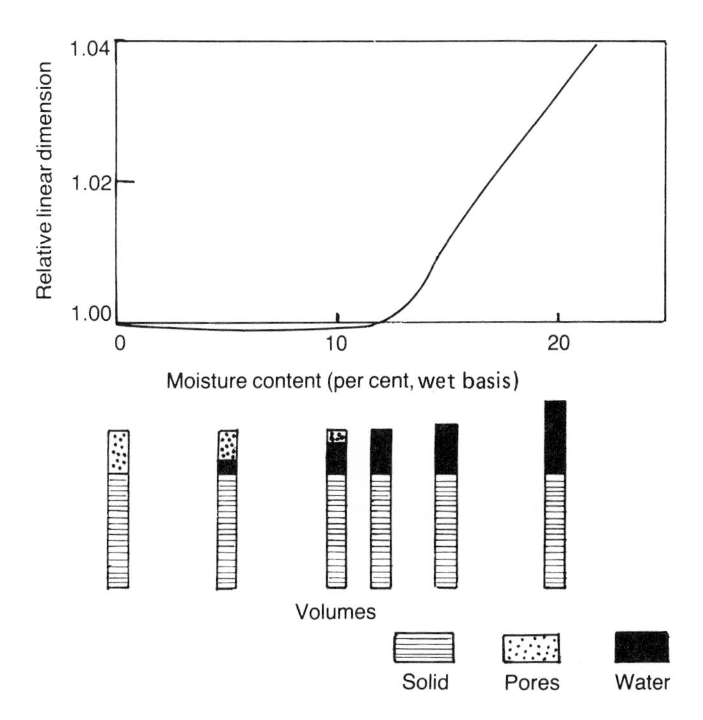


Fig. 10.4 – Size change during drying.

operate, pulling the particles closer together, and thus producing a further small contraction. Different sized pores empty at different times, so there is no sharp change in the curve. However, as more and more pores become empty, the surface tension forces diminish and the particles separate slightly, resulting in a small expansion to the point of complete dryness.

In the case of powder pressed articles, such as wall tiles, there is an additional complication in the elastic recovery after pressing. In addition to the expansion towards the zero moisture content, there is another slight contraction as the last traces of moisture are removed. These size changes are small, but they are sufficient to account for some types of cracking.

## 10.7 ANISOTROPY

We have so far considered only isotropic contraction, but in practice most pottery bodies show marked anisotropy in their drying behaviour. This is mainly due to the preferred orientation of the clay particles, imposed by the fabrication process. In both casting and plastic forming, the clay platelets tend to line up parallel to each other in a particular direction. As a consequence, there is a greater number of interparticle water layers contained in unit distance in the direction perpendicular to the plates, compared with those in a perpendicular direction. Thus, when the water is removed the contraction is correspondingly greater, and may often be twice as great in the normal as in the parallel direction. In general, high clay content in the body leads to greater anisotropy. This can be an important additional factor to add to the presence of moisture gradients when considering cracking or warping.

## 10.8 CRACKING AND WARPING

Cracking can occur when the strain resulting from size differentials exceeds a critical value. Consider the drying of a circular plate set in a bung. If drying takes place from the rim of the plate, large differences in moisture content can develop between the edge and the centre. The edge seeks to contract, but cannot do so because the centre has not contracted as much. So the rim is in tension, and a crack may develop such as is shown in Fig. 10.5. This may happen over a range of moisture contents. In quantitative terms, although the strength and modulus of elasticity vary markedly with moisture content, rupture usually occurs at a constant value of the strain, at approximately 0.1 per cent. Measurements on earthenware type bodies show that moisture content differences of 2 or 3 per cent can produce strains of 0.1 per cent. In the case of powder pressed bodies, such as tiles, there is an interesting consequence to the contraction phenomenon in that it sometimes happens that an edge crack will appear early in the drying, then close up when the article is dry, and reappear during the firing.

Even when cracking does not occur, the problem of warpage during drying can be troublesome. The principal point to note here is that differential shrinkages set up due to moisture gradients are frequently not recoverable when the moisture

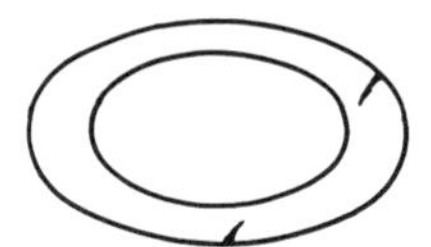

Fig. 10.5 – Cracking of a plate rim.

contents are equalized. A simple example is shown in Fig. 10.6, which shows a rectangular slab of body. At stage a, the moisture contents are uniform, and there are no stresses. If the slab is dried from the top surface only, the top will contract, as in stage b. The bottom part, being still capable of plastic flow, takes up the new shape without undue strain. If the whole piece is now dried the bottom surface seeks to contract but cannot do so because the top has become rigid. So the final shape is as shown in stage c. If the differences are too great cracking will occur as well as warpage. The position becomes more complicated if one part of the article is restrained and another part is free to move, as is frequently the case in practice when articles have to be loaded on top of each other. Changes in shape and thickness within a piece also tend to accentuate the problem. The only answer is to control the rate of drying so that the moisture gradients are kept within the required limits.

## 10.9 SENSITIVITY

Tendency for cracking to occur is a function of body characteristics as well as of drying conditions. In the same environment some bodies crack and other do not. Such factors as composition, and especially clay content, particle size and pore distribution, unfired strength, together with rather more abstruse parameters such as moisture adsorption, and aqueous conductivity, all have their relevance. Attempts have been made to construct from these parameters a so-called 'sensi-

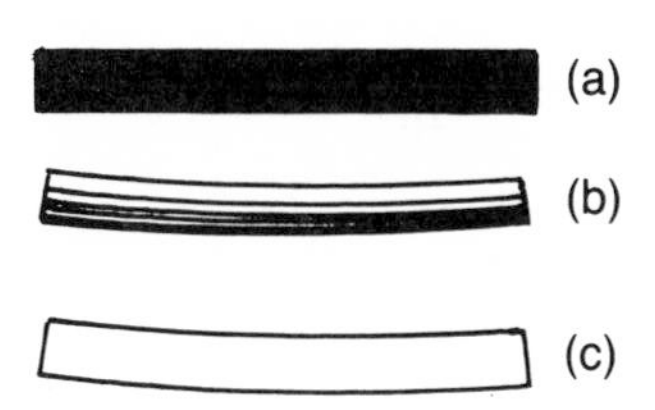


Fig. 10.6 – Warpage of a slab on drying.

tivity index', which would attribute to a body a number indicating its inherent tendency to crack. It has not been possible to find a single index giving an unambiguous pointer to this characteristic (Hermann, 1966; Hermansson and Carlsson, 1979).

However, some success has been achieved on an empirical basis. Various experimental techniques have been used. Generally a specially shaped test piece is subjected to a controlled drying regime such that there is a good probability of cracking taking place. Sometimes the test piece is constrained in such a manner as to prevent some size changes taking place. Observation of the times, moisture contents, moisture gradients, size changes, and other factors, at the point when cracking occurs enables an index to be determined that places different bodies in a reasonable order of merit. With such a complex phenomenon, much more than this is not likely to be achieved.

## REFERENCES

Ford, R. W. (1964) *Institute of Ceramics Testbook,* 3. Drying.
Hermann, R. (1966) *Tr. Int. Cer. Congress. Stockholm,* 79.
Hermansson, L. and Carlsson, R. (1979) *Tr. & J. Brit. Cer. Soc.,* **78**(5), 98.
Moore, F. (1961) *Tr. Brit. Cer. Soc.,* **60**, 517.

# 11

# The effect of heat

When ceramic bodies are heated to temperatures of 1000° C or more they undergo permanent, irreversible change. Indeed, this is almost the way in which ceramics are defined, the Greek word *Keramos* meaning burnt material. In this connection ceramics may be distinguished from glasses in that they are first shaped and then fired. The changes in whiteware bodies are related mainly to the effects produced by the application of heat to clay, so it is necessary to consider these in some detail.

## 11.1 THE EFFECT OF HEAT ON CLAY

In noting the effect of heat on clay, it is convenient to look first at kaolinite, since that is the main constituent of most clays in industrial use. It needs to be remembered, however, that, although the main features are to be found in other clays, there are significant differences, for example, with halloysite and montmorillonite. Also, real clays are seldom pure, and some impurities can introduce modifying characteristics. Notwithstanding these reservations, the behaviour of kaolinite under heat treatment can be taken as sufficiently representative to give a sound basis to the understanding of firing technology. Dry clay has a fair measure of strength and rigidity, but if it is re-wetted it can be returned to a plastic or slip condition. Fired clay, on the other hand, loses its plasticity permanently, and does not return to its original state when water is added. It acquires a new level of strength and chemical durability. What happens when heat is applied?

Observation by simple experimental techniques yields interesting information. Even in the earliest days Josiah Wedgwood observed that a loss of weight occurs when clay is heated. More detailed study yields a curve of weight loss against temperature, the sample being held for a sufficient time at each temperature for a steady state to be attained. A typical result for an air-dry kaolinite is shown in Fig. 11.1(a). There is a small weight loss, of the order of 1 per cent, up to about 150 °C, representing the removal of the remaining mechanical water, since clay in equilibrium with the atmosphere adsorbs some moisture and stabilizes at a finite but small moisture content. In the temperature range 400–600 °C, however, a much larger loss occurs, of the order of 14 per cent, and it can easily be shown that the matter being lost is water. From there on to the peak heating temperature, which may be in the range 1100 to 1250 °C, the weight is substantially constant, though there are small changes due to the burning out of combustible matter. Thermal expansion is another property that is easily measured,

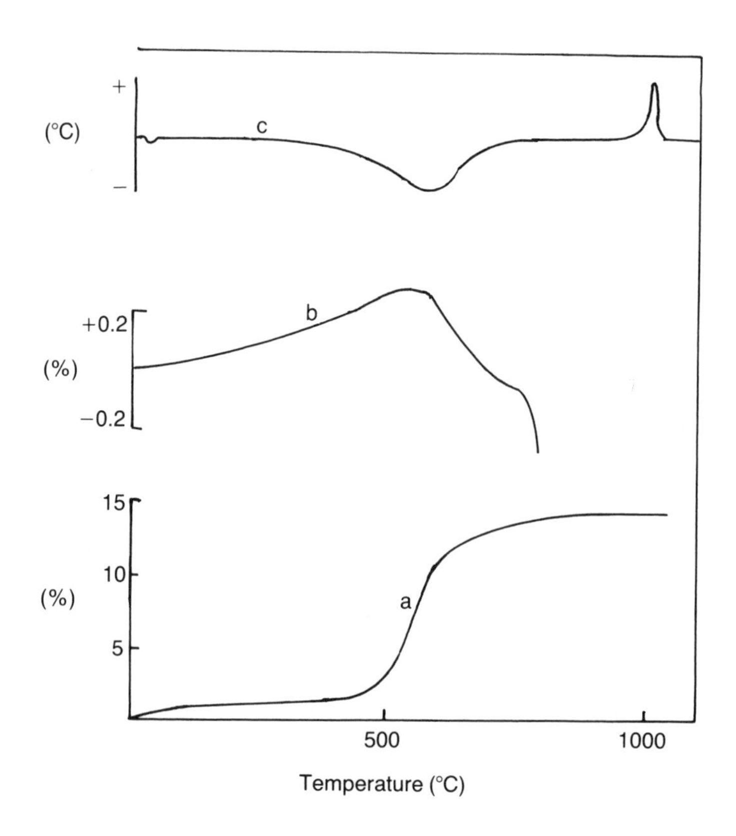


Fig. 11.1 – Effect of heat on clay. a, weight loss; b, thermal expansion; c DTA.

and a typical curve for kaolinite is shown in Fig. 11.1(b). It will be seen that there is a steady expansion up to about 500 °C, when a rapid contraction occurs, again indicating that some fundamental change is taking place.

A technique that has yielded much useful information in recent years is that known as differential thermal analysis, commonly referred to as DTA. In this method, a sample under test is heated alongside a sample of similar size, but known to be inert, and the temperature difference between the two is observed. Whenever reactions involving the evolution or absorption of heat take place, the curve shows a peak – above the reference line for exothermic reactions, and below for endothermic. If the conditions, and especially the rate of heating, are standardized, the position and magnitude of these peaks are found to be characteristic of particular minerals. Much useful information about mineral composition and structural changes can thus be obtained. Such a DTA pattern for kaolinite is shown in Fig. 11.1(c). Three features can be noted. First, there is a small endothermic reaction at the beginning, corresponding to the removal of the mechanical water already mentioned in connection with weight loss. Then there is a major change taking place, again in the 500 to 600 °C range, and involving the take-up of heat. Finally, approaching 1000 °C, there is a sharp exothermic peak.

All these observed phenomena are now well understood, in the light of the information yielded on crystal structure by x-ray studies. The major change in the 500 to 600 °C range is associated with the breakdown of the kaolinite lattice and the release of the combined water. Following this, so-called metakaolin is formed, followed by spinel, which eventually leads to the formation of mullite. The rejected silica eventually melts to form a glass. There has been much debate about the nature of the exothermic reaction at 980 °C, but it is generally agreed that it is associated with the spinel and mullite formation. Strangely, although the lattice is completely broken down, kaolinite crystals can often be seen to retain their characteristic shape over the whole of this temperature range.

The weight loss in the 500 to 600 °C region, can be understood in the light of the known composition of the kaolinite molecules. The reaction is as follows:

| | Kaolinite | Metakaolin | | Water |
|---|---|---|---|---|
| | $Al_2O_3.2SiO_2.2H_2O$ | $Al_2O_3.2SiO_2$ | + | $2H_2O$ |
| Molecular weight | 258 | 222 | | 36 |

The weight loss is thus 36 ÷ 258 = 13.95 per cent, corresponding to the value shown in Fig. 11.1(a). As a result of the loss of this water, the kaolinite is irreversibly changed – in the sense that it has lost its plasticity and strength, the latter only being regained by reaction with other body materials at higher temperatures.

## 11.2 THERMAL DATA

The thermal data associated with the structural changes are not so easily determined, and the published values show wide variations. For present purposes, however, the quantities of heat involved may be taken to be roughly as shown in Table 11.1.

**Table 11.1**

| | Quantity of heat (MJ $kg^{-1}$) |
|---|---|
| Removal of mechanical water per unit weight of water | 2.76 |
| Removal of combined water per unit weight of kaolinite | 630 |
| Exothermic reactions at 1000 °C per unit weight of kaolinite | 125 |

The other data needed for a quantitative appraisal of the firing process are the interval specific heats of fired and unfired body materials. The published values for these also show some variation, but practical values to 1100 °C can be taken as

| | |
|---|---|
| Unfired clay substance | 2.1 kJ $kg^{-1}$ $°C^{-1}$ |
| Fired clay and other body materials | 1.4 kJ $kg^{-1}$ $°C^{-1}$ |

From these data, some assessment can be made of the basic heat requirement in the firing of a typical earthenware body.

Consider 1 kg of unfired body, and suppose the body to contain 35 per cent clay, 65 per cent non-clay material and to hold 2 per cent mechanical water at the start of firing; and suppose the firing to be taken to a maximum temperature of 1100 °C. Then the various elements of the heat requirement are as follows:

(1) To heat up and evaporate 0.02 kg of water requires
$$0.02 \times 2.76 \times 10^3 = 55 \text{ kJ}$$
(2) To heat up 0.343 kg of clay from 20 to 1100 °C requires
$$0.343 \times 2.1 \times 1080 = 778 \text{ kJ}$$
(3) To heat up 0.637 kg of non-clay material from 20 to 1100 °C requires
$$0.637 \times 1.4 \times 1080 = 963 \text{ kJ}$$
Thus the total heat needed is 1796 kJ.

The weight of the combined water would be $0.343 \times 0.1395 = 0.048$ kg so that the weight of the fired body would be $1 - 0.02 - 0.048 = 0.932$ kg. If this fired body were cooled down to 20° C, the amount of heat given up would be

$0.932 \times 1.4 \times 1080 = 1409$ kJ.

The amount of irrecoverable heat used in the firing of the dry body is thus $1741 - 1409 = 332$ kJ. We may summarize these data (in round figures) as follows.

| | kJ | Per cent |
|---|---|---|
| Total heat supplied | 1796 | 100.0 |
| Heat used for drying | 55 | 3.0 |
| Irrecoverable heat | 332 | 18.5 |
| Heat recoverable on cooling | 1409 | 78.5 |

It is thus seen that of the heat supplied just over 20 per cent is used up in the firing process. Given a suitable firing system some of the remainder of the heat can be recovered, and the significance of this in the assessment of kiln efficiency will become apparent in the next chapter.

The foregoing reasoning requires some further comment. The numerical values will be different for bodies that differ in composition from the earthenware example taken, but the basic principle remains valid. In calculating weight loss, account needs to be taken of the burning off of organic matter, and the loss associated with the breakdown of other compounds. The heat contributed by the organic matter is generally of the order of 80 kJ per kg of body. With some clays, however, it may be as high as 400 kJ, which represents a substantial contribution to the heat requirement.

## 11.3 TEXTURAL CHANGES

The structural and constitutional changes caused by the application of heat, bring about changes in texture. Perhaps the most important of these is the change in porosity accompanying the formation of glass in the body. Porosity measurements give a very useful indication of the degree of vitrification, and they are also directly indicative of volume shrinkage.

The porosity changes occurring during the firing of typical earthenware and bone china are shown in Fig. 11.2. We need here to make a distinction between true and apparent porosity. Apparent porosity is the pore volume accessible to fluids under normal conditions of measurement; given sufficiently hard firing it can usually be reduced to zero. Even at this stage, however, most bodies still contain anything between 5 and 10 per cent by volume of closed pores. The true porosity is defined as the sum of the open and closed porosity. At high porosities, all the pores tend to be accessible, so that the true and apparent porosities have the same value.

Referring to Fig. 11.2, it will be seen that the earthenware type of body increases in porosity up to about 1000 °C. However, some textural changes are taking place; some of the smaller particles are dissolving, and there is a marked decrease in the internal surface area at this stage. There is also an increase in pore size. Beyond this temperature, vitrification proceeds and above 1100°C the porosity decreases rapidly, coming to zero at about 1220 °C. The slope of this part of the curve is an important characteristic of the body system, as it determines the firing range. The true porosity does not reach zero, and at about 1250 °C the pores begin to expand again (so-called bloating), and the pore volume increases. Beyond this point the apparent porosity also begins to increase, but the body is now seriously overfired and there are bound to be problems with distortion and other faults.

In the case of bone china, the body starts with a higher porosity which increases slightly at first, then remains fairly constant until rapid vitrification sets in at a higher temperature than in the case of earthenware.

The concept of firing range merits a more detailed explanation. In practical firing circumstances, there are always differences in temperature from point to point in the kiln, and even at different points in one piece of ware. It is important to know what range of associated porosity differences can be tolerated; either in further processes such as glazing and decorating, or in the properties of the finished product. A useful method of considering the firing range problem, is to use bulk density as the parameter, rather than porosity. The two are related by the relation

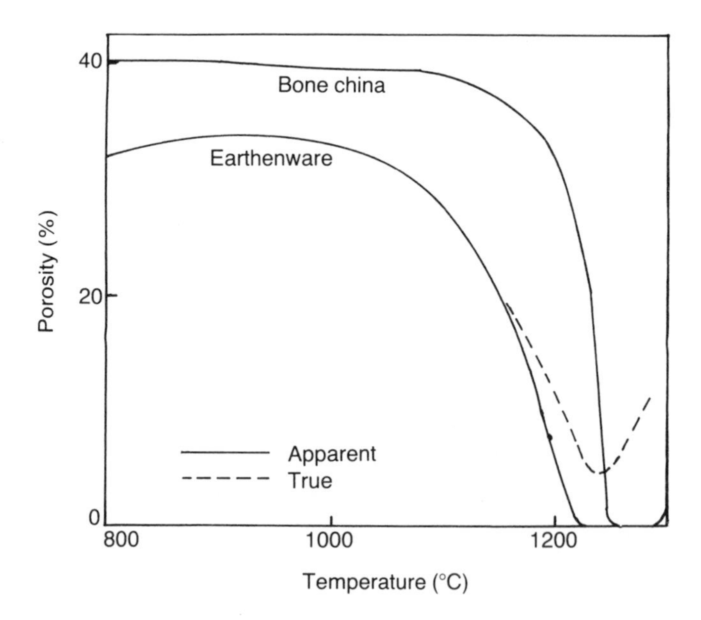


Fig. 11.2 – Porosity changes during firing.

$$\text{Bulk density} = \text{true density of solid } (1 - \text{porosity})$$

Assuming a value for the solid density, the earthenware firing curve shown in Fig. 11.2, has been replotted in terms of bulk density in Fig. 11.3. Given prescribed limits of bulk density, this curve serves to indicate what temperature differences can be tolerated. It also indicates what happens to the firing range when changes are made to the body, for example in composition or in size of materials.

Norris *et al.* (1979) have pointed out that changes in bulk density are accompanied by changes in many other properties, and have developed the concept of range curves as a means of identifying acceptable firing conditions. For example, strength is closely related to porosity, and passes through a maximum as the temperature is increased. However, this maximum does not usually occur at the same temperature as that of the bulk density, so a choice may have to be made as to whether the product demands optimum strength or maximum bulk density. Other properties such as translucency and thermal expansion do not show a maximum, but increase or decrease continuously beyond the peak in bulk density. The choice of an optimum firing schedule is thus a matter for

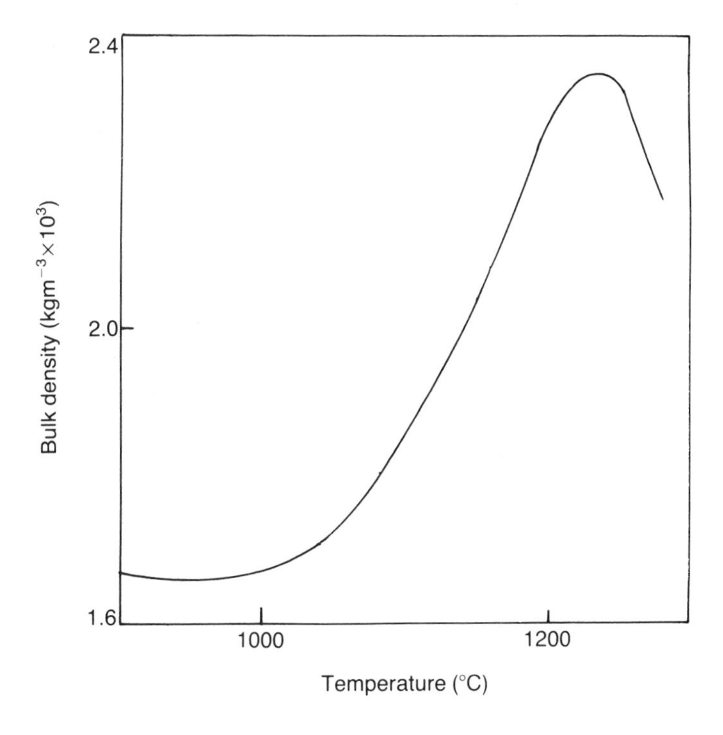


Fig. 11.3 – Bulk density change during firing.

compromise, optimizing as many significant properties as possible.

At higher temperatures, deformation becomes critically important. Increasing glass formation, and a lower viscosity, lead to a state in which the body deforms, either under gravity or under the influence of imposed stresses or differential contraction. The problem is especially important in the case of vitrified products. Since the process of vitrification can only take place through glass flow, it usually happens that as the lower values of porosity are reached, the tendency to deform begins to manifest itself. This is one of the main reasons why control of firing temperature is of crucial importance.

## 11.4 TEMPERATURE AND TIME

A phrase often used in connection with ceramic firing is 'heat work'. This is an ill-defined term, but it is intended to convey the concept of total energy input, which may take the form of a low temperature for a long time, or a high temperature for a short time. There are clearly limits at both extremities; there is a minimum temperature below which the reactions will not proceed however long the time, and there is an upper limit to the temperature that the ware will withstand. In between, there is presumed to be some degree of interchangeability between temperature and time; and this has practical significance in relation to such questions as very rapid firing. It is, therefore, of some interest to consider whether any such relationship exists. Taylor (1979) has examined the vitrification process in bone china, including isothermal data, and has shown graphically how temperature and time are related for different degrees of vitrification. His treatment is closely followed here.

A convenient index to take is the porosity, indicating as it does the degree of vitrification. The curve of porosity against temperature shown in Fig. 11.2, represents what happens during a firing in which the temperature is continuously increased up to a maximum. In order to analyse the effect of time it is necessary to have isothermal data; that is, measurements of the porosity over a period of time at a series of fixed temperatures. Some data exist in the literature, but, of course, the numerical values are different for different bodies. For instance, the reactions taking place in bone china are quite different from those in the clay–quartz–felspar system. Nevertheless, there is enough evidence to suggest that, at least in the later stages of vitrification, we are concerned with a first order reaction. If we assume that this is so we can proceed with a possible line of approach.

Let $p_0$ be the initial porosity and $p$ the porosity after time $t$. If we have a first order reaction, $\log (p/p_0) = -kt$. Taking an arbitrary time for a given reduction in porosity, a value of $k$ can be established. A curve of porosity against time can then be calculated, of the form shown in Fig. 11.4.

The constant $k$, a measure of the reaction rate, is, of course, a function of the temperature. This relationship is of the form $k = Ae^{-B/T}$ or $\log_e k = \log_e A - B/T$, where $T$ is the absolute temperature, and $A$ and $B$ are constants. Assuming values for $A$ and $B$, a family of porosity–time curves for a range of

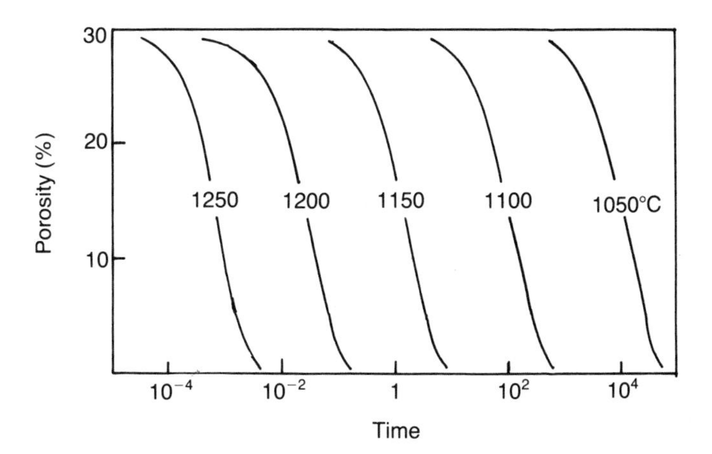


Fig. 11.4 – Temperature–time relationship for vitrification.

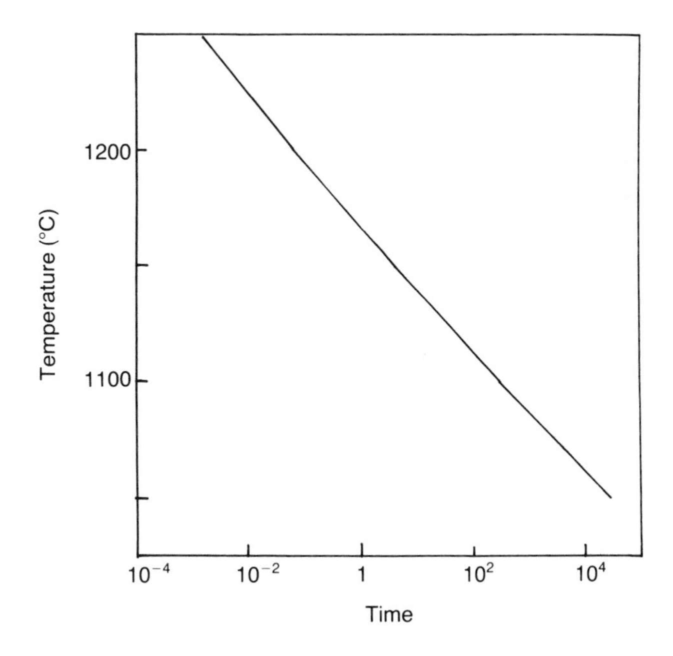


Fig. 11.5 – Temperature–time relationship for porosity reduction from 30 to 5 per cent.

temperatures can be calculated, as shown in Fig. 11.4. Choosing a particular degree of vitrification, a relationship between time and temperature can be read off. Such a curve is shown in Fig. 11.5 for a reduction in porosity from 30 to 5 per cent. This curve shows that a tenfold change in time is associated with a temperature difference of about 30 °C at the higher temperatures, and about 25 °C at the lower end. In spite of the differences between body systems it is often found that the temperature difference for a tenfold time change, does fall between 20 ° and 50 °C.

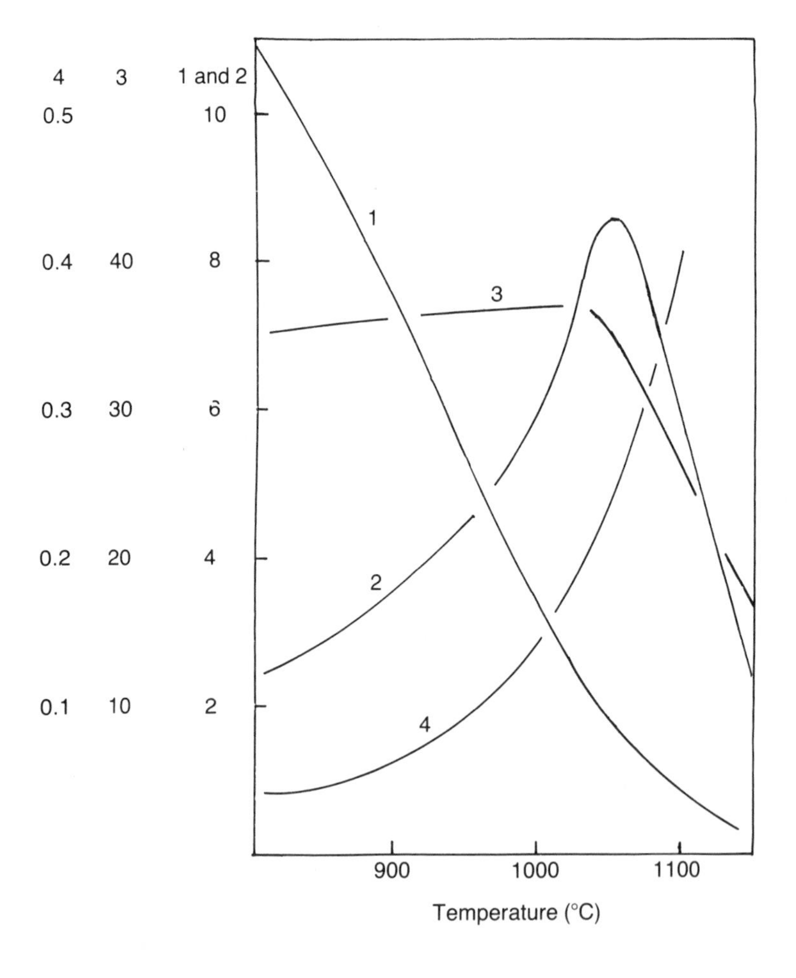


Fig. 11.6 – Textural features of an earthenware body. 1, Specific internal surface ($m^2\ kg^{-1} \times 10^3$). 2, air permeability ($cm^2\ s^{-1}\ cm\ Hg^{-1} \times 10^{-4}$). 3, apparent porosity (%). 4, mean pore radius ($\mu m$).

## 11.5 PERMEABILITY

The understanding of changes in texture during firing is assisted by a number of measurable properties, of which porosity, bulk density, internal surface area and pore size, have already been mentioned. Another property that is an indicator of pore texture is the permeability. This is low, even for gases, but it can be measured, and it supplements the data from other techniques. Given certain assumptions it can be related to pore size and surface area. In a wide pore-size distribution, it is mainly influenced by the larger pores, unlike surface area which reflects the influence of the finer pores.

When all these physical parameters are collected together, as is shown in Fig. 11.6 for an earthenware body, a fairly complete picture of the effect of heat on texture emerges (Binns and Dinsdale, 1959). As a result of dehydration and the combustion of organic matter, the porosity increases slightly up to about 1000 °C, after which there is a rapid fall as vitrification sets in. Long before this there is a sharp decrease in internal surface area, as small particles dissolve and small pores amalgamate to form larger ones. The surface area continues to decrease throughout the firing. As the pore size increases, so does the permeability, until the vitrification takes over, so that the permeability curve shows a pronounced maximum. Pore size values deduced from data on permeability and porosity, on surface area and porosity, and measured by the mercury penetration

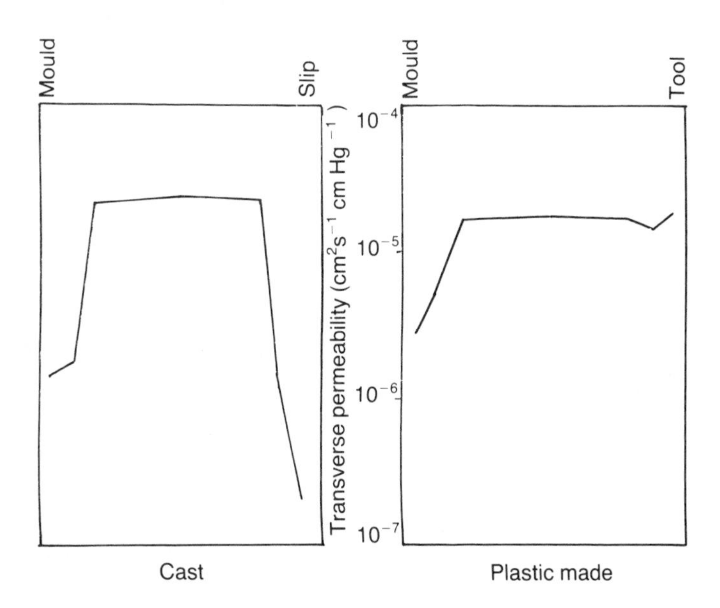


Fig. 11.7 – Variation in transverse permeability through the thickness of fired ware.

method, all show reasonable agreement.

The fact that permeability, unlike the other data, has a directional aspect makes it a useful tool for studying anisotropy in texture. For example, a plate made by either casting or a plastic process, will show permeability values along the principal plane several times greater than those in a direction perpendicular to this, i.e. through the thickness. These differences do not appear in powder-pressed articles, such as tiles. The technique is also a sensitive index of changes in texture throughout the thickness of a piece of ware. A typical set of such transverse permeability data are shown in Fig. 11.7, for cast and plastic made earthenware. In both cases, there is a thin impermeable layer, developed on firing, near the surface in contact with plaster. A similar layer occurs on the slip side on the cast article; this may be due to the migration of alkalis to that surface on drying giving rise to increased vitrification. In the case of plastic ware, there is usually very little variation in permeability on the tool side.

## 11.6 ATMOSPHERE

Some of the reactions taking place in the body on heating are influenced by the nature of the surrounding atmosphere. Normally, the kiln atmosphere should be oxidizing, in order that combustible matter may be burnt out. In the case of hard porcelain, however, the atmosphere is delberately starved of oxygen, in order to develop the desired body properties. Fired colour is particularly susceptible to atmosphere, since different colours are produced by different states of oxidation of metallic constituents, such as iron.

Water vapour can also have a marked effect. By reducing the viscosity of the glassy component, it markedly increases the vitrification rate. It also accelerates the conversion of quartz to cristobalite, thus increasing the thermal expansion at high temperatures.

It is also now known that other elements in the atmosphere can play an important part. Fluorine is an interesting example. Fluorine compounds from either the clay or the flux can be found in kiln atmospheres. When they react with the exposed surfaces of ware, they can reduce the vitrification and the contraction. This can result in surface porosity, and in distortion of the ware (Holmes, 1973).

## 11.7 COOLING

The final stages of firing are important in that if the rate of cooling is too high, cracking of the fired article will occur. The main reason for this is that the outside of the piece, for example, the rim of a plate, will be at a lower temperature than the centre, thus resulting in a tangential stress at the rim. If the critical strain for the fired ware is exceeded cracking occurs.

Internal flaws can also be produced by rapid cooling through the quartz and cristobalite inversion points.

## REFERENCES

Binns, D. B. and Dinsdale, A. (1959) *The A. T. Green Book,* 277. (British Ceramic Research Association).

Holmes, W. H. (1973) *Tr. & J. Brit. Cer. Soc.,* **72**(1), 25.

Norris, A. W., Taylor, D. and Thorpe, I. (1979) *Tr. & J. Brit. Cer. Soc.,* **78**(5), 102.

Taylor, D. (1979) *Tr. & J. Brit. Cer. Soc.,* **78**(2), 43.

# 12

# Firing

The firing process has always been a central element in ceramic technology. The effectiveness of the operation depends on the development of the required properties in the finished product. Further significance is added in present times by the emphasis on the efficient use of energy.

In this chapter we shall be concerned only with the so-called biscuit fire, in which the clay based body is transformed by heat into a strong rigid material capable of being glazed and decorated. Many of the general considerations apply also to the glaze firing, but the special features of glazes will be dealt with in Chapter 13.

## 12.1 FIRING SCHEDULE

The firing process consists essentially of increasing the temperature of the ware progressively over a period of time, holding it at a peak temperature, and then cooling back to room temperature. The rate at which the temperature can safely be changed varies at different stages, and there is thus a characteristic shape to a firing curve. The absolute values of the permissible rates, and the total time involved vary over wide limits according to the type of ware, the type of kiln, and the bulk and density of the setting. Nevertheless, we can identify a number of principles that are generally applicable.

### 12.1.1 Drying

It might be thought that since most products undergo a specific drying process, no drying would be needed in the kiln. However, there is nearly always a small amount of residual pore water and, in addition, physically adsorbed water on the internal surfaces. This water will be driven off in the kiln over the temperature range up to 250 °C, and the rate of heating over this period must be low enough to avoid explosive vaporization in the ware. Allowing time for the water vapour to diffuse through the ware is particularly important in the case of thick articles, such as sanitary ware, or a dense setting, such as a bung of tiles or plates.

### 12.1.2 Dehydroxylation and oxidation

From 250 °C to about 500 °C the heating can proceed fairly rapidly but at that temperature two important reactions occur that require more time. The first is the removal of the combined water from the clay molecule. This represents about 5 per cent by weight in many bodies. In other words in a large kiln containing twenty tons of ware, a ton of water has to be evaporated and removed.

At about the same time, any carbonaceous matter present in the ware is burnt out; here, time is required for the oxygen and the products of combustion to diffuse through the ware. This diffusion is not helped by the fact that the movement of water vapour is taking place at the same time.

### 12.1.3 Vitrification

From about 850 °C onwards to the onset of cooling is commonly known as the vitrification or finishing period. In this range the fluxes melt, a glassy phase develops, and appropriate crystalline changes occur. In order that such fired properties as strength or translucency may develop adequately there usually has to be a period when the ware is held at the peak temperature for a time – often known as the 'soaking' period. The peak temperature may be as low as 1050 °C for some clay bodies, generally in the region of 1150 °C for earthenware, 1250 °C for bone china or porcelain, and up to 1450 °C for hard paste porcelain. Hard porcelain is a special case, in which a preliminary low temperature biscuit fire is followed by a high temperature glaze fire incorporating a period with a reducing kiln atmosphere.

### 12.1.4 Cooling

Once the appropriate degree of vitrification has been achieved, the ware can be cooled, usually at a rate much faster than the heating rate. Even so, care has to be taken to avoid excessive temperature gradients, or cracking or dunting may be encountered.

### 12.1.5 Crystalline inversions

In both heating and cooling the rate needs to be slowed down when passing through crystalline inversions involving volume changes. Those of most importance are the cristobalite inversion in the range 200–250 °C, and the quartz inversion in the range 550–600 °C.

Bearing all these considerations in mind, it is possible to build up a notional ideal firing schedule, and some of the important features will usually be found in practice. For example, Fig. 12.1 shows the mean temperature as measured by a travelling thermocouple placed in the setting in a tunnel kiln firing biscuit wall tiles.

However, there are several reasons why it is not possible to achieve the ideal curve for all parts of the ware in industrial conditions. In the older type of coal-fired intermittent ovens, big variations occurred when coal was added to the fires, the temperature of the ware sometimes actually falling back as the fires die down. Similar irregularities are sometimes found when the ware passes burners in gas-fired tunnels, or heating elements in electric.

Then there are inevitable variations between different parts of the setting. In large settings, temperature differences of several hundred degrees can often develop between different parts, so that some time has to be spent in allowing equalization to take place.

The size of the setting makes a big difference to the possible rate of firing, and dense settings such as tiles or flat tableware need longer times than open settings. Thus, times to maximum temperature can vary from 50 to 100 hours in large intermittent ovens, and from 20 to 100 hours in traditional tunnel kilns. In modern small intermittent kilns complete firing cycles within 24 hours are possible. With very small settings, much faster times can be achieved; the subject of fast firing is of great topical interest, and will be dealt with in Section 12.10.

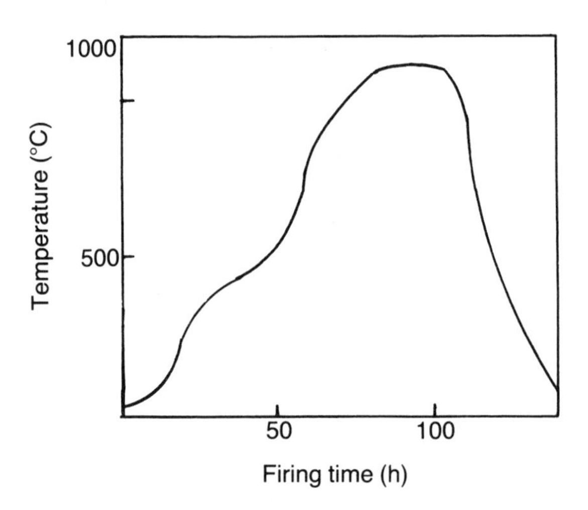


Fig. 12.1 – Biscuit firing schedule for wall tiles.

## 12.2 TYPES OF KILNS

Heat treatment of ware in practice requires the use of some kind of kiln or oven, designed to enable the appropriate temperature-time schedule to be achieved. There are clearly two quite distinct ways in which this can be done. One is to use a structure in which the ware is stationary and the temperature is progressively varied; the other is to set up a kiln in which there is a temperature distribution that varies in space but is constant in time, and through which the ware is caused to move. These are usually known as intermittent and tunnel kilns respectively, and it is convenient to consider the basic principles of kiln design under these two headings.

### 12.2.1 Intermittent

Large coal-fired intermittent ovens have now been largely replaced by tunnel ovens. Nevertheless, we set down here for the record some aspects of this type of firing that are of interest. Some of the principles may be relevant to the operation of the new generation of smaller intermittent units. Dinsdale (1953) reviewed the historical development of different types of firing, and more recent reviews have been published by Holmes (1967, 1978).

The intermittent oven is one in which the ware is placed, either open or in some form of refractory container, the oven then being heated up and cooled down over a period time. The heat may be provided by any one of a number of sources, but the traditional method was by burning coal in a series of firemouths situated around the perimeter. The oven is usually constructed of firebrick and contained in an outer brick structure, the function of which is to provide a chimney. This outer hovel, as it is often called, is traditionally built in a tapering shape; it is often known as a bottle oven, and is a familiar sight in a pottery-producing area. The operation of the oven requires considerable skill on the part of the fireman, and part of this skill is in manipulating the combustion and draught conditions in such a way as to follow the correct temperature requirements, and also to keep all parts of the oven sharing a similar heat treatment. Since some of these ovens may be 5 to 10 metres in diameter, the latter is not easily achieved.

Control of the flow of gases is crucial to the successful operation of these ovens, and many systems of draught distribution have been developed. These are variations on two main principles, known as up-draught and down-draught, shown in vertical section in Fig. 12.2.

The up-draught oven has a series of firemouths, perhaps twelve, around the circumference in which the coal is burnt. The hot gases enter the oven through an aperture known as a bag, the bag wall protecting the ware placed close to the firemouth. A proportion of the hot gases travel along radial underfloor flues, entering the oven at offtakes along the radius and at the centre. This assists the even distribution of heat, and minimizes the tendency for a cold region to develop at the centre bottom of the setting. The pressure at the top of the oven is less than that at the firemouths, so there is a continuous upward flow, regulated by a central damper in the crown and a number of smaller quarter dampers. The draught at the firemouth can also be reduced by the use of regulator holes above the bag; other holes, known as spy-holes, are used for visual inspection of the setting.

One of the disadvantages of this system is that the spaces between bungs of ware act as chimneys, so that when a channel becomes overheated, the gas flow is increased; to that extent the system is unstable. This tendency can be checked by the use of the quarter dampers, which also enable different quarters of the setting to be kept at the same temperature. During the early stages of firing the top dampers are open, to allow the moisture from the ware to escape, and the hot gases pass straight up through the bag, with very little penetration along the underfloor flues. At this stage the ware at the top of the setting and that nearest the walls gains in temperature much more rapidly than the rest. As time goes on the pressure at the top of the oven increases, the neutral pressure line descends, and is usually held by the fireman at about the level of the spy-hole. Later in the firing, although the top of the setting will still be ahead, there will have been some conduction along the floor. The regulator slides are then brought into play and have the double effect of holding back the top by allowing cold air to enter, and reducing the draught at the bag, so causing more gases to be pulled along the underfloor flues to the centre of the oven.

In the down-draught oven, offtakes are provided in the floor so that hot gases can be drawn off through cavities in the wall of the oven. This improves the heat distribution, and also improves the thermal efficiency by cutting down

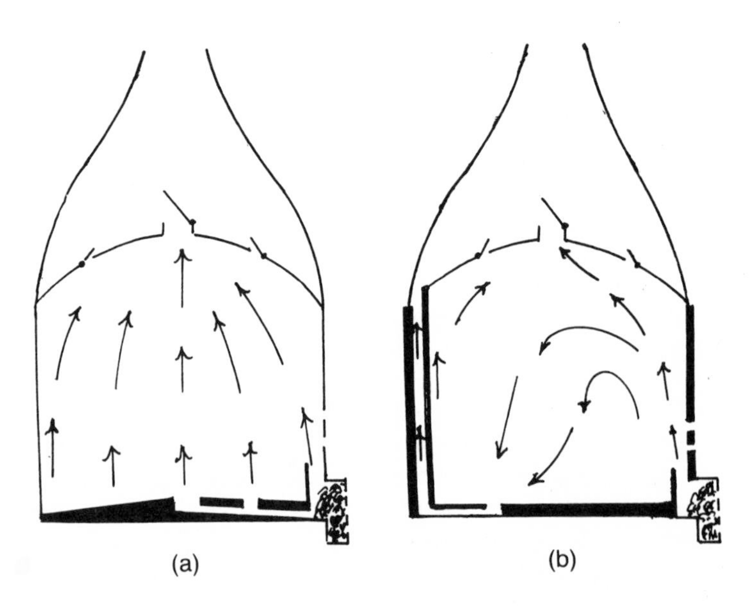


Fig. 12.2 – Vertical section through intermittent ovens. (a) Updraught (b) Downdraught.

Fig. 12.3 – First five tunnel kiln for once-fired dinner ware. (Courtesy Drayton Kiln Co.).

some of the heat losses through the walls. There are many combinations of the two systems, perhaps best described as up-and-down draught ovens. In recent years there has been considerable advance in the use of intermittent kilns of much smaller dimension, generally carrying the setting on a single truck that can be pushed in and out of the kiln. The availability of ceramic fibre insulation has enabled lightweight structures to be used. The development of rapid firing has meant that in many cases the firing cycle can be completed within 24 hours, this making for a very convenient production routine. Advantages claimed for this type of kiln are improved heat distribution, better thermal efficiency, faster firing and more flexible working schedules. They may be heated by oil, gas, or electricity.

### 12.2.2 Continuous

In continuous firing systems, the temperature distribution in the kiln is static and the ware is moving (see Fig. 12.3). The tunnel may be of the order of 100 metres long, and the ware is carried on trucks and insulated bases and running on rails. If the heat is supplied by electricity, or by gas or oil confined to muffle chambers, the setting may be placed open on kiln furniture. If open-flame conditions prevail, the ware generally, though not invariably, has to be placed in saggars.

In contrast to the intermittent ovens, the tunnel oven is recuperative in principle, the hot gases and the ware moving in opposite directions. The plan diagram, Fig. 12.4, shows that there are three main zones, for preheating, firing, and cooling respectively. The hot gases are drawn through the kiln and are taken off at the main exhaust. On the way, they give up a large proportion of their heat to the incoming ware, which is thereby preheated. The products of combustion thus leave the kiln at a very much lower temperature than is the case with the intermittent process. After the ware has passed through the firing zone, it enters the cooling zone, which often consists of a hollow wall and roof construction, through which cooling air is drawn. Some of this cooling air, having taken up heat from the ware, can be fed back to the burners as preheated

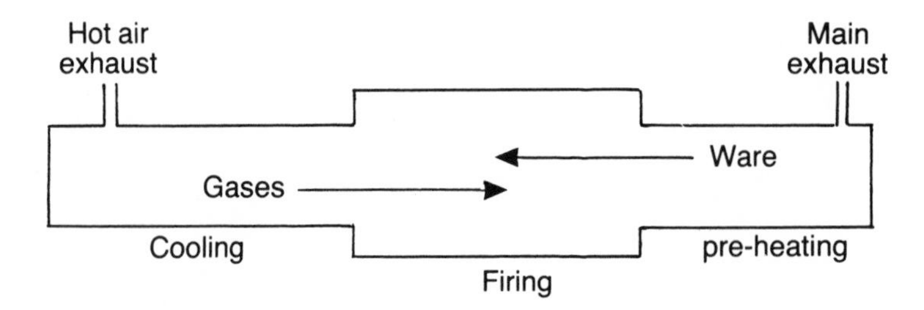


Fig. 12.4 – Plan of tunnel oven system.

combustion air. The system is not thermodynamically balanced, as there is usually an excess of clean hot air which is available for use for purposes external to the kiln, such as drying or space heating.

Many variations in design have emerged over the years, including kilns in which the ware is carried on a roller hearth, multi-passage kilns with counter-flow, and so on, but the general principle of continuous motion remains the same. Because of this, the tunnel kiln can be so placed as to fit in with a planned continuous flow of ware through the factory.

## 12.3 HEAT TRANSFER

The heating of ware in a kiln involves all three of the classical modes of heat transfer, namely, radiation, conduction, and convection. The relative importance of these varies according to the type of kiln, the fuel used, and the furnace design. For example, in the traditional coal-fired oven, at least in the early stages, convection is mostly responsible for the heat transfer; on the other hand, where electrical heating is used, radiation plays a predominant part.

Radiation from a source of heat increases in importance as the temperature is increased. The total energy emitted is proportional to the fourth power of the absolute temperature. It also depends on the emissivity of the radiating surface, and its effectiveness depends on the emissivity of the receiving surface, which is a measure of the capacity to absorb radiation. Emissivity data are notoriously difficult to come by, and for a given material are greatly influenced by the smoothness of the surface and the colour, as well as being temperature dependent. For ceramic materials the value may vary from 0.5 to as high as 0.95. In the practical situation radiation may be from electric heating elements, flames, brickwork surfaces, tubes containing combustion gases, burner blocks with combustion in the pores, and so on. The radiation from flames can be increased by so arranging the combustion conditions that there are luminous particles present. One of the disadvantageous features of radiation is that it is directional, and thus an article requiring to be heated may be in the shadow of another, so that it would be almost impossible to achieve uniform heating in a bulk setting by radiation alone.

In these circumstances most kilns rely heavily on heat transfer by convection. Transport of heat by moving gas is a very effective method of achieving temperature uniformity. The gas may be in the form of the hot combustion products from fossil fuels, or in the form of hot air specially circulated for the purpose. Since heat transfer increases with the velocity of the gas, many devices have been developed for encouraging more rapid movement, but the need in many cases to avoid excessive dust disturbance places a limit on what can be done.

The third process, conduction, is always significant. It helps to distribute heat throughout the furnace structure, at the same time unfortunately being responsible for heat loss through the walls and roof. The coefficient of thermal conductivity is defined as the amount of heat passing through unit area in unit time through unit thickness of a substance when there is unit difference in temperature between the two faces. Some typical approximate values given below show that ceramic materials lie somewhere between metals and some organic substances.

| | $J\ m^{-1}\ s^{-1}\ K^{-1}$ |
|---|---|
| Wood | 0.15 |
| Plastic | 0.20 |
| Water | 0.60 |
| Ceramic | 1.20 |
| Iron | 80 |
| Copper | 380 |

The conductivity varies considerably with composition, texture, and temperature. Conduction through the kiln furniture is an important element in heat transfer to the ware. Thus the conductivity of alumina and silicon carbide are, respectively, about 1½ and 20 times that of firebrick.

In achieving uniformity of temperature throughout a piece of ware, thermal diffusivity is of crucial importance. Heat transfer through walls when a steady state prevails is easily calculated from thermal conductivity alone. However, in the unsteady state it is often desirable to know the way in which temperature is changing, rather than how much energy is being transferred. In this case the key parameter is the thermal diffusivity, which is defined as the thermal conductivity, divided by the product of the specific heat and the density. For pottery bodies it has a value of approximately $0.03 \times 10^{-4}\ m^2\ s^{-1}$ in the unfired state, and about double this when fired. Since the diffusivity varies with temperature, the solution of the differential equation for unsteady heat flow is difficult. However, numerical solutions can be worked out showing how the temperature distribution changes with time. The details of this analysis are beyond the scope of this book, but its importance is clear when we remember that pottery bodies are virtually opaque to radiation. What happens in the firing process is that heat is transferred by radiation and convection, and some conduction, to the surface only of a piece of ware. From then on the process is entirely controlled by conduction within the ware itself. The thermal diffusivity determines the temperature differences that can arise between different parts of an article, and hence gives an indication of safe rates of heating. The importance of this in studies of possible very fast firing rates for single pieces of ware is clear. The possibility of rapid firing in relation to these parameters has been examined by Holmes (1969).

## 12.4 INSULATION

Heat loss through furnace walls is unavoidable and is of serious concern in firing economics. A standard method of reducing it is to increase wall thickness, and to use composite wall structures, with high temperature firebrick at the operating surface, backed by a layer of insulating brick. Since the conductivity of insulating brick may be only one-tenth of that of firebrick, it clearly saves unacceptably large increases in wall thickness. All insulation involves capital expenditure, and this has to be balanced against energy saving in any particular situation.

A very significant development in recent years has been the manufacture of ceramic fibre and associated components. This has revolutionized possibilities in kiln design. Lightweight structures, having a combination of low conductivity and low thermal mass, are increasingly coming into use, especially in intermittent kilns.

## 12.5 THERMAL EFFICIENCY

The definition and measurement of thermal efficiency in kilns is a very large and complex subject, and can only be treated in outline here. Since the concepts are fundamentally different for intermittent and continuous firing it is necessary to consider the two separately.

### 12.5.1 Intermittent firing

In the intermittent firing process, the whole of the kiln structure, as well as the ware, is heated up and cooled back to room temperature in each firing cycle. Thus, in addition to the losses through the structure as firing proceeds, all the sensible heat in the structure is lost at the end. A very simple index of the efficiency of the furnace is the energy consumption per unit weight of ware. This figure may vary from up to 80 MJ $kg^{-1}$ in coal-fired bottle ovens, 4–20 in tunnel ovens, and as low as 2 in multipassage or special intermittent kilns. All these figures are much higher than that required to bring about constitutional changes in the ware. In order to have a better understanding of the efficiency of the process it is necessary to have information on the heat losses, so that a heat balance can be drawn up. Consider the following data, which are fictitious but which could be typical of a coal-fired bottle oven firing tableware.

| | |
|---|---|
| Weight of fired ware | $3 \times 10^4$ kg |
| Weight of remainder of setting | $5.6 \times 10^4$ kg |
| Total heat supplied from fuel to oven | 630 GJ |
| Heat to fire ware to peak temperature | 51 GJ |
| Heat to fire remainder of setting to peak temperature | 69 GJ |
| Recoverable heat in ware at peak temperature | 37 GJ |
| Recoverable heat in remainder of setting at peak temperature | 69 GJ |
| Heat lost in flue gases | 300 GJ |
| Heat lost through kiln structure and on cooling | 210 GJ |

From these data we can construct a heat balance account, as follows, the heat units being in GJ.

| | |
|---|---|
| Heat supplied 630 (100%) | |
| Used in firing ware | 14 (2.2%) |
| Lost in cooling total setting | 106 (16.8%) |
| Lost in flue gases | 300 (47.7%) |
| Lost by radiation and convection and cooling of kiln structure | 210 (33.3%) |

Such a balance sheet indicates how the heat is used or dissipated, and provides a useful guide to the most rewarding areas for improvement.

As an index of the effectiveness of the firing operation, we see that the fuel consumption is 21 MJ $kg^{-1}$ for the ware, and 11.25 MJ $kg^{-1}$ for the total setting. Clearly, any improvement in the ratio of ware to setting will improve the consumption figure for the ware.

If we define the thermal efficiency of the operation as the ratio of the heat usefully employed to the total heat input, we have

Thermal efficiency for ware = 51 ÷ 630 = 8 per cent
Thermal efficiency for total setting = 120 ÷ 630 = 19 per cent

These simple parameters yield useful comparisons as between different types of kilns all operating on the intermittent principle, with little or no heat recovery. However, the efficiency difinition is not applicable to the situation in continuous kilns without some qualification.

### 12.5.2 Continuous firing

A typical set of data, for a period of one hour, for a gas-fired tunnel oven firing tableware might read as follows:

| | |
|---|---|
| Weight of fired ware | 200 kg |
| Weight of remainder of setting | 400 kg |
| Total heat supplied | 2700 MJ |
| Heat to fire ware to peak temperature | 340 MJ |
| Heat to fire remainder of setting to peak temperature | 510 MJ |
| Irrecoverable heat used in firing ware | 105 MJ |
| Fuel consumption for ware | 13.5 MJ $kg^{-1}$ |
| Fuel consumption for total setting | 6.75 MJ $kg^{-1}$ |

If we use the same definition of thermal efficiency as we used for intermittent kilns, we have

Thermal efficiency for ware = 340 ÷ 2700 = 12.6 per cent
Thermal efficiency for total setting = 850 ÷ 2700 = 31.5 per cent

It is clear that if the heat transfer from flue gases to ware is very effective, it is possible for the efficency figure to exceed 100 per cent. However, the fuel consumption per unit weight is still a valid index of the thermal perfcrmance of the kiln.

A heat balance account can be constructed, and for this kiln might be somewhat as follows, the heat units now being in MJ per hour.

| | |
|---|---|
| Heat supplied 2700 (100%) | |
| Irrecoverable heat used in firing ware | 105 (3.9%) |
| Sensible heat in setting leaving kiln | 65 (2.4%) |
| Sensible heat in car structure leaving kiln | 65 (2.4%) |
| Sensible heat in flue gases | 840 (31.1%) |
| Sensible heat in cooling air | 590 (21.9%) |
| Radiation and convection loss from kiln structure | 1035 (38.3%) |

## 12.6 FACTORS AFFECTING THERMAL EFFICIENCY

Consideration of the foregoing analysis of heat utilization leads to the conclusion that thermal efficiency cannot be regarded as a constant characteristic of a kiln design. It is a function of a number of important factors in the operating conditions, as well as being dependent on size. It is useful to look at three of these factors in more detail.

### 12.6.1 Throughput

The amount of ware passing through a tunnel kiln in a given time can easily be altered by changing the car travel speed. In practice nearly all manufacturers are tempted to increase the throughput above that for which the kiln was nominally designed, and it is often found that this can be done without any adverse effect on the quality of the ware. The result is invariably an improvement in thermal performance, a consequence that can easily be theoretically predicted from a study of the heat balance. Consider the gas-fired kiln, the heat balance for which is shown above, and suppose that when the throughput is raised by a fraction $\delta t$ the gas consumption increases by a fraction $\delta g$. The amount of heat lost from the kiln structure will remain the same, namely 1035. The flue gases heat loss 840 will increase in proportion to the gas consumption. All the other elements, 105 + 65 + 65 + 590 = 825, will increase in proportion to the increase in the throughput. Thus,

$$1035 + 840\,(1 + \delta g) + 825\,(1 + \delta t) = 2700\,(1 + \delta g)$$

from which

$$\frac{\delta g}{\delta t} = 0.44$$

Hence, the increase in gas consumption is only about 44 per cent of the increase in throughput, with a corresponding increase in the thermal efficiency. For various reasons, the improvement in practice is not as good as this, the figure for a range of kilns being around 30 per cent rather than 44 per cent.

### 12.6.2 Size

Since a substantial proportion of the heat lost from a tunnel kiln is by way of radiation and convection from the surface, and since the ratio of surface to volume decreases with increased size, it might be expected that large tunnel kilns would be more efficient than small ones. However, another factor to be taken into account is the variation of firing time with cross-sectional area. There are reasons to think that the thermal efficiency, $E$, might be related to the square root of the cross-sectional area, $x$, by an equation of the form

$$E = C x^{1-\rho}$$

where $C$ is a constant and $\rho$ is a measure of the way in which the firing time, $t$, varies with the size, assuming $t = kx^{\rho}$. The value of $\rho$ determines whether the efficiency increases with size or not. For $\rho > 1$ the efficiency increases with size; for $\rho < 1$, the reverse is the case; for $\rho = 1$, efficiency is independent of size. Considerations of thermal conductivity would predict $\rho = 2$ for a solid setting, and results for closely-packed brick setting confirm this. In the case of pottery, with a much more open setting, $\rho$ lies between 1 and 2, possibly of the order of 1.2.

### 12.6.3 Temperature

Temperature-time characteristics have an influence on efficiency. Clearly, if the setting is held for some time at the peak temperature, more heat will be required to make up for the increased radiation loss, but the amount of heat used in firing to the peak is the same. Thus the nominal thermal efficiency is unaltered, but the true efficiency is less. From a thermodynamic point of view, heat cannot be regarded as an exchangeable quantity regardless of temperature. A true heat balance would be better drawn up in terms of entropy, and if this were done the effect of the temperature-time cycle on efficiency would be more clearly understood.

## 12.7 UNIFORMITY OF FIRING

One of the problems of firing is that of achieving uniformity of heat treatment throughout the setting. When the temperature is rising, thermal gradients are inevitable, and it is not always possible to even them out by holding the temperature for a period. The aim must always be to confine the heat treatment variations within the firing range for the body concerned, so that all products have acceptable properties.

A typical set of contraction trial ring values indicating the amount of heat work, for a typical up-draught coal-fired oven firing tableware, would show that the hardest-fired part of the setting is just above the firemouths, and that there are easy-fired regions in the centre, at both top and bottom.

Similar differences are encountered in tunnel kilns, but the situation is more complex here in the sense that the range is materially altered when the car speed is changed. In Fig. 12.5, the trial ring values for a setting of wall tiles are shown for different car speeds. As the speed is increased, the mean ring value is reduced and the range is increased. The effect is clearly shown in the table below.

| *Rate of travel (cars/day)* | *Average ring value* | *Percentage of setting within a range of 5 points* |
|---|---|---|
| 11 | 42½ | 100 |
| 15 | 37½ | 63 |
| 23 | 35½ | 48 |

Measurement of temperature in various parts of the setting, by means of thermocouples for example, can assist in locating the source of the worst variations. Improvements can be made by increasing the circulation of the kiln atmosphere, shielding the setting from direct radiation, supplying extra heat to the cooler regions, and other devices.

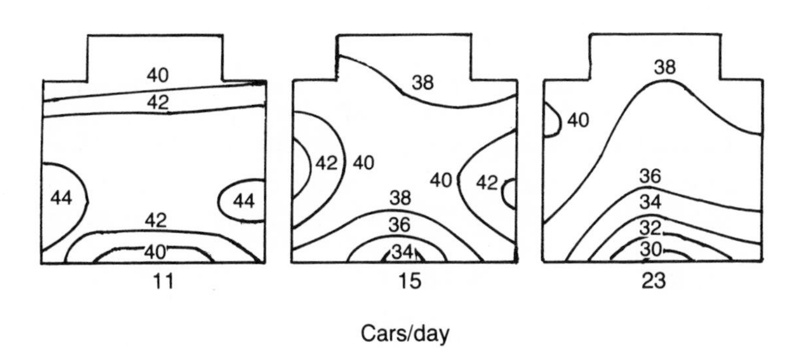


Fig. 12.5 – Distribution of contraction ring values in a wall-tile tunnel oven.

## 12.8 MEASUREMENT AND CONTROL

The refinement of control techniques that has developed over the years has stemmed mainly from the introduction of improved measuring techniques. The parameters for which it is desirable to have quantitative data include temperature, heat work, atmospheric composition and gas flow.

Traditional coal-fired intermittent ovens were fired without the use of precise temperature measurement. Use was made of cones, fabricated from different mixtures of a vitrifiable body. When these were set in the kiln at an angle to the vertical, they bent over as successively higher temperatures were reached. They were observed through spy-holes in the kiln wall. As an additional measure of heat work, contraction trial rings were placed throughout the setting, including places from which they could be drawn as firing proceeded. The rings were made from body material, and the heat work was represented by the amount of contraction that had taken place, the fired diameter being easily measured on a pointer gauge. Attempts were made to construct tables, attributing temperatures to particular ring values, but these were not very accurate.

The increasing availability of thermocouples enabled actual temperatures to be measured at fixed points; optical or radiation pyrometers could also be used for the same purpose. These readings, however, did not give a true or complete picture of the condition of the ware. This was eventually achieved by placing the thermocouples in the setting. In the case of tunnel kilns, they were arranged so as to travel through the kiln with the cars. In this way, it is now possible to obtain a complete record of the true firing schedule undergone by different parts of the setting.

Similarly, it is now possible to extract gas samples continuously, and record the more important aspects of the composition of the kiln atmosphere. Physical techniques for determining oxygen and carbon dioxide contents have greatly facilitated this process.

Given this data, and the use of computerized control systems, substantial improvements in firing control, and in achieving a desired result, can be obtained.

## 12.9 SOURCES OF HEAT

Nearly every available source of energy can be used in the firing process. The use of coal has declined over the years, owing to many factors including clean air legislation. Of the others, natural gas, LPG, oil and electricity, all have their advocates. In these days of high energy costs, thermal efficiency has a special significance. Ease of control is also an important factor. Relative costs per unit of heat tend to change over the years, often being at the mercy of fiscal policies. For this reason, many modern kilns are designed in such a way as to facilitate changes from one fuel to another.

## 12.10 RAPID FIRING

Much interest has been shown in recent years in the possibility of improving kiln performance by a substantial reduction in firing time (Holmes, 1969). It is now well established that single articles can be satisfactorily fired in extremely short times, provided that the temperature is increased and that uniformity of temperature across the article can be established. Most of the required physical properties can be achieved, though some, such as translucency, would seem to need time for their full development.

It follows that the possibility exists for the use of kilns firing a layer of articles one high, on a suitable continuous conveyor such as a belt or roller hearth. Simple shapes, such as tiles, would seem to offer the best prospects of success. Results from systems in industrial production do not yet seem to present a clear picture as to whether there are substantial savings in energy compared with traditional kilns. The question is complicated by many other factors, such as capital cost, total output requirements, space considerations, integration with other production elements, and so on. Undoubtedly the development will continue, and rapid firing is certain to find successful application in particular circumstances.

## 12.11 GLOST AND ONCE-FIRED WARE

This chapter has been almost wholly concerned with various aspects of biscuit firing. Most of the principles outlined are also applicable in the case of glazed ware, but there are some special problems in the maturing of glazes that will be dealt with in a later chapter.

## REFERENCES

Dinsdale, A. (1953) *British Ceramic Society. A Symposium,* 363.
Holmes, W. H. (1967) *J. Brit. Cer. Soc.,* **4**(3), 289.
Holmes, W. H. (1969) *J. Brit. Cer. Soc.,* **6**(2), 19.
Holmes, W. H. (1978) *Tr. & J. Brit. Cer. Soc.,* **77**(1), 25.

# 13

# Glazes

Some ceramic products fulfil their function well enough without glazing. Fine-grained, well vitrified bodies, such as jasper, basalt, or parian, can be formed into artware so as to give an aesthetically attractive appearance without any additional surface layer. For the most part, however, the advantages conferred by the addition of a glaze layer are substantial, and for this reason glazing has become an integral part of whiteware technology.

There are many aspects to be noted. The smooth, glassy layer, which is usually of the order of 100 $\mu$m in thickness, provides a high degree of gloss when required, but can equally well produce attractive non-glossy effects, as in matt or satin finishes. It provides an impermeable layer on porous bodies. It is easy to clean, an important aspect in the case of wall tiles or sanitary ware. It protects underglaze decoration, and forms an ideal base on which to apply on-glaze decoration. The glaze itself can provide decoration, either by the use of coloured glazes *per se* or in terms of so-called in-glaze decoration. The use of coloured glazes goes far back in history, probably to the Egyptians, and is responsible for much of the charm of early pottery. It is an extraordinary fact that many wonderful coloured glazes were produced without the benefit of the knowledge of modern theories of glass structure.

In addition to aesthetic effects, glazes have important technical characteristics. Thus, they can materially increase mechanical strength, or provide specific electrical properties.

The details of glaze formulation are very complex, and constitute a special

Fig. 13.1 – Automatic glaze spraying line for flatware. (Courtesy Service Engineers.)

study that is well beyond the scope of this book. For this reason, also, most pottery manufacturers do not compound their own glazes, but take them from specialist glaze suppliers. What we seek to do in this chapter is to set out some of the physical principles involved in glaze formulation, application, maturing, and influence on properties.

## 13.1 APPLICATION

Glazes may be applied under a variety of circumstances. They may be applied to unfired bodies in the so-called once-fired systems; to partially fired bodies, as in the case of hard-paste procelain; or, as is most common, to fired biscuit ware.

The glaze is usually in the form of a suspension in water. It may be applied by dipping the article in a tub of glaze; by spraying in atomized form; or, as in the case of wall-tiles, by means of a waterfall. All these processes are becoming increasingly mechanized (Fig. 13.1). In some special cases, a vapour deposition technique may be used, this being the modern equivalent of the old salt-glazing.

The physical properties of the glaze slip are of crucial importance in ensuring the deposition of a uniform layer of the correct thickness over all parts of the article. Among the more important requirements are:

(1) The ability to remain in suspension.
(2) Appropriate rheological properties, giving good flow but a high yield value, to avoid running down vertical surfaces.
(3) Good unfired strength, adhesion, and low contraction.

In order to meet these requirements, compositions have to be carefully balanced, particle size distribution has to be controlled, and sometimes organic additives have to be used.

It is not often realized that in terms of surface tension the unfired layer is in unstable equilibrium with the body surface, especially as the thickness is generally less than the critical thickness needed to ensure spreading. Incorrect glaze compounding, or contamination of the biscuit surface, can give rise to changes in surface energy which result in the fault commonly known as crawling, in which the glaze layer separates into islands, leaving exposed areas of biscuit that do not heal on firing (Budworth, 1971).

## 13.2 COMPOSITION

Glazes are fundamentally similar to glasses, though often containing more components and melting at appreciably lower temperatures. All glazes contain silica, which itself melts to form a glass at around 1700 °C. If certain oxides are mixed with silica, the melting point is reduced; in some cases mixtures of oxides form a eutectic with a melting point lower than that of either component. In terms of modern theories of glass structure, silica may be regarded as a network former, and these fluxing oxides as network modifiers. They break the silicon-oxygen bonds, and lower the melting temperatures, to around 1000 °C.

It is common practice in setting out glaze compositions to group the network formers on one side, the network modifiers on the other side, with oxides of an intermediate nature in between. In other terms, these oxides can be classified as acidic, basic, and amphoteric. The principal oxides used in glazes, and grouped in this manner are shown below.

| *Basic* | *Amphoteric* | *Acidic* |
|---|---|---|
| $Na_2O$ | | |
| $K_2O$ | | |
| $CaO$ | $Al_2O_3$ | $SiO_2$ |
| $MgO$ | | $B_2O_3$ |
| $PbO$ | | |
| $SnO$ | | |
| $ZnO$ | | |

The list is by no means exhaustive, many other oxides being added for special purposes. When the constituent oxides of a glaze are expressed in molecular proportions, it is customary to reduce the formula so that the sum of the basic oxides is 1.

Some of the oxides can be added in oxide form, but others are derived from complex oxide minerals. Thus clay will contribute $Al_2O_3$ and $SiO_2$; flint or quartz, $SiO_2$; felspar $K_2O$ and $Na_2O$, together with $SiO_2$ and $Al_2O_3$; whiting $CaO$; litharge $PbO$; talc $MgO$ and $SiO_2$; borax $B_2O_3$ and $Na_2O$, and so on. In our considerations we shall regard these as pure minerals, whereas in practice many of these raw materials contribute traces of many oxides other than those for which they are principally used.

These raw materials cannot be used in their primitive form, except in very limited art work. Two considerations impose certain requirements. In the first place, some of the oxides are soluble in water, so they cannot be used in glaze slips until they have first been converted into an insoluble form. Secondly, some materials, such as lead oxide, are toxic and have to be converted into a harmless form; for example lead bisilicate, $PbO.2SiO_2$, is non-toxic. These transformations are brought about by a preliminary melting of some of the glaze components to form a frit, which is chilled and ground to a powder. The two principal frits in common use are lead bisilicate frit and borax frit. Either or both of these may be added to the remainder of the batch, known as the milled materials, to form the complete glaze.

## 13.3 CALCULATIONS

Glaze compositions can be expressed in either molecular proportions, or percentages by weight. We illustrate the calculations involved in deducing batch composition from glaze composition; the reverse process is similar in principle. The reservation should always be kept in mind that the composition of a thin layer of glaze after firing will differ from the nominal as a result of reaction with the body and volatilization of some of the constituents.

The molecular weight data for the oxides involved are as follows:

| | | | | | |
|---|---|---|---|---|---|
| $SiO_2$ | 60.1 | $Al_2O_3$ | 102.0 | CaO | 56.1 |
| $K_2O$ | 94.2 | $Na_2O$ | 62.0 | PbO | 223.2 |
| $B_2O_3$ | 69.6 | $CO_2$ | 44.0 | $H_2O$ | 18.0 |

For our example, we take a low-solubility lead glaze to be compounded from a lead bisilicate frit, a borax frit, and a mill batch. The required glaze composition is, in, molecular proportions,

$$\left.\begin{array}{ll} \text{CaO} & 0.4 \\ \text{K}_2\text{O} & 0.1 \\ \text{Na}_2\text{O} & 0.3 \\ \text{PbO} & 0.2 \end{array}\right\} 1.0 : \quad \text{Al}_2\text{O}_3\ 0.35 : \quad \begin{array}{ll} \text{SiO}_2 & 3.0 \\ \text{B}_2\text{O}_3 & 0.4 \end{array}$$

Multiplying these molecular proportions by the appropriate molecular weights we arrive at the composition by weight shown in Table 13.3.

We consider first the use of a low solubility lead frit, based chiefly on lead bisilicate, $PbO.2SiO_2$, and having the following molecular composition:

$$\left.\begin{array}{ll} \text{CaO} & 0.06 \\ \text{K}_2\text{O} & 0.02 \\ \text{PbO} & 0.92 \end{array}\right\} : \text{Al}_2\text{O}_3\ 0.1 : \text{SiO}_2 \quad 1.85$$

The possible way in which this frit could be made up from the available raw materials is set out in Table 13.1.

**Table 13.1** – Lead bisilicate frit – molecular proportions

| | CaO | $K_2O$ | PbO | $Al_2O_3$ | $SiO_2$ |
|---|---|---|---|---|---|
| | 0.06 | 0.02 | 0.92 | 0.10 | 1.85 |
| 0.92 mol litharge | | | 0.92 | | |
| | 0.06 | 0.02 | | 0.10 | 1.85 |
| 0.06 mol whiting | 0.06 | | | | |
| | | 0.02 | | 0.10 | 1.85 |
| 0.02 mol potash felspar | | 0.02 | | 0.02 | 0.12 |
| | | | | 0.08 | 1.73 |
| 0.08 mol clay | | | | 0.08 | 0.16 |
| | | | | | 1.57 |
| 1.57 mol flint | | | | | 1.57 |

On the first horizontal line we set out the required molecular proportions of the oxides. We first provide for the lead, by supplying 0.92 mol litharge (PbO). After subtracting this, we have a new line of requirements. The CaO can be provided by 0.06 mol of whiting ($CaCO_3$). We note here that when we calculate the weights in the batch, we need to allow for the loss of $CO_2$. Next we add some felspar. This could be in the form of potash felspar ($K_2O.Al_2O_3.6SiO_2$); 0.02 mol. of this provide the necessary $K_2O$, but also contribute some $Al_2O_3$ and $SiO_2$. If a mineral such as cornish stone were used, there would be different proportions of alkalis, alumina, and silica. We now need to provide some $Al_2O_3$ and $SiO_2$, and this could come from 0.08 mol. of clay. Finally, the required 1.57 mol. of $SiO_2$ could come from flint or quartz.

We can then work out the batch requirement as follows.

| *Material* | *Formula* | *Mol. prop.* | *Mol. wt.* | *Weight* | *Percentage weight* |
|---|---|---|---|---|---|
| Litharge | PbO | 0.92 | 223.2 | 205.3 | 60.8 |
| Whiting | $CaCO_3$ | 0.06 | 100.1 | 6.0 | 1.8 |
| Potash felspar | $K_2O.Al_2O_3.6SiO_2$ | 0.02 | 556.8 | 11.1 | 3.3 |
| Clay | $Al_2O_3.2SiO_2.2H_2O$ | 0.08 | 258.2 | 20.6 | 6.1 |
| Flint | $SiO_2$ | 1.57 | 60.1 | 94.4 | 28.0 |

Consider next the borax frit, with the molecular composition:

$$\left.\begin{array}{ll} CaO & 0.45 \\ K_2O & 0.10 \\ Na_2O & 0.45 \end{array}\right\} : Al_2O_3 \ 0.3 : \begin{array}{ll} SiO_2 & 3.5 \\ B_2O_3 & 0.9 \end{array}$$

Following the same line of reasoning as for the lead frit we can meet the requirements by using borax, whiting, felspar, clay, and flint. The batch composition is as follows

| *Material* | *Formula* | *Mol. prop.* | *Mol. wt.* | *Weight* | *Percentage weight* |
|---|---|---|---|---|---|
| Borax | $Na_2O.2B_2O_3.10H_2O$ | 0.45 | 381.2 | 171.5 | 36.2 |
| Whiting | $CaCO_3$ | 0.45 | 100.1 | 45.0 | 9.5 |
| Potash felspar | $K_2O.Al_2O_3.6SiO_2$ | 0.10 | 556.8 | 55.7 | 11.7 |
| Clay | $Al_2O_3.2SiO_2.2H_2O$ | 0.20 | 258.2 | 51.6 | 10.9 |
| Flint | $SiO_2$ | 2.50 | 60.1 | 150.2 | 31.7 |

We now need to calculate the proportions of these two frits required, and the batch composition needed to provide the remainder of the glaze. In Table 13.2 the first line shows the required glaze composition. The $B_2O_3$ will be

**Table 13.2** – Glaze composition – molecular proportions

| | CaO | $K_2O$ | $Na_2O$ | PbO | $Al_2O_3$ | $B_2O_3$ | $SiO_2$ |
|---|---|---|---|---|---|---|---|
| Glaze | 0.40 | 0.10 | 0.30 | 0.20 | 0.35 | 0.40 | 3.00 |
| Borax frit | 0.20 | 0.044 | 0.20 | | 0.133 | 0.40 | 1.555 |
| Lead frit | 0.013 | 0.004 | | 0.20 | 0.022 | | 0.402 |
| Remainder | 0.187 | 0.052 | 0.10 | | 0.195 | | 1.043 |
| 0.187 mol whiting | 0.187 | | | | | | |
| | | 0.052 | 0.10 | | 0.195 | | 1.043 |
| 0.052 mol potash felspar | | 0.052 | | | 0.052 | | 0.306 |
| | | | 0.10 | | 0.143 | | 0.737 |
| 0.1 mol soda felspar | | | 0.10 | | 0.100 | | 0.600 |
| | | | | | 0.043 | | 0.137 |
| 0.043 mol clay | | | | | 0.043 | | 0.086 |
| | | | | | | | 0.051 |
| 0.051 mol flint | | | | | | | 0.051 |

supplied by the borax frit. The amount required is 0.40; so that the amount of frit needed is 0.4 ÷ 0.9, and reducing the borax frit composition by this fraction, we obtain the contribution from the borax frit, as shown in the second line. Similar reasoning for the lead frit, using the fraction 0.2 ÷ 0.92, gives the contribution shown in the third line. Adding the two frits, and substracting the total from the glaze, we obtain the required contribution from the mill batch, shown in the fourth line. This could be provided by whiting, potash felspar, soda flespar, clay and flint, as shown in Table. 13.2. We thus arrive at the following batch composition for the non-frit materials.

| *Material* | *Formula* | *Mol. prop.* | *Mol. wt.* | *Weight* | *Percentage weight* |
|---|---|---|---|---|---|
| Whiting | $CaCO_3$ | 0.187 | 100.1 | 18.72 | 16.4 |
| Potash felspar | $K_2O.Al_2O_3.6SiO_2$ | 0.052 | 556.8 | 28.95 | 25.3 |
| Soda felspar | $Na_2O.Al_2O_3.6SiO_2$ | 0.100 | 524.6 | 52.46 | 45.9 |
| Clay | $Al_2O_3.2SiO_2.2H_2O$ | 0.043 | 258.2 | 11.10 | 9.7 |
| Flint | $SiO_2$ | 0.051 | 60.1 | 3.06 | 2.7 |

Therefore the three parts of the glaze need to be mixed in proportions relating to their molecular weights, thus:

| | Weight | Weight per cent Unfired | Fired |
|---|---|---|---|
| Lead frit | 72.17 | 20.69 | 21.28 |
| Borax frit | 162.74 | 46.65 | 48.00 |
| Milled material | 113.94 | 32.66 | 30.72 |

It will be noted that the milled material percentage in the unfired state is higher than in the fired state to allow for the loss of water from the clay and $CO_2$ from the whiting.

All the above analysis has been carried out in terms of molecular formulae, since this gives a good picture of the glaze composition. It could equally well be done in terms of weight percentages. The results of such a calculation are shown in Table 13.3.

**Table 13.3** – Glaze compounding in weight proportions

| Oxide | Lead frit | Borax frit | Milled batch | Glaze |
|---|---|---|---|---|
| CaO | 1.01 | 6.89 | 10.07 | 6.6 |
| $K_2O$ | 0.57 | 2.57 | 4.70 | 2.8 |
| $Na_2O$ | | 7.62 | 5.95 | 5.5 |
| PbO | 61.86 | | | 13.2 |
| $Al_2O_3$ | 3.07 | 8.36 | 19.10 | 10.5 |
| $B_2O_3$ | | 17.11 | | 8.2 |
| $SiO_2$ | 33.49 | 57.45 | 60.18 | 53.2 |
| Per cent in glaze, by weight | 21.28 | 48.00 | 30.72 | 100 |

## 13.4 TYPES OF GLAZES

Many different types of glazes are needed to meet the wide range of requirements in use. They are nearly all based on a silicate composition, but with a wide range of compositions. Leadless glazes, firing at temperatures as high as 1450 °C, are commonly used on hard-paste porcelains; at somewhat lower temperatures they are appropriate for sanitary ware. For earthenware and bone china, low-solubility lead glazes are used, fusing in the 1000-1100 °C range. High lead glazes on artware fuse at still lower temperatures. Many artistic effects are achieved by departing from the clear, smooth, transparent system. A small degree of vitrification gives a satin surface; larger crystals produce a matt effect. Opacity can be obtained by the incorporation of such components as tin oxide or zirconia, which stay undissolved in the glaze and act as light-scattering particles. Coloured glazes are produced by the inclusion of colouring oxides; random mixed coloured effects can be produced by the application of coloured glazes in two immiscible liquids. Sometimes glazes are required to have specific physical properties. Thus glazes with a significant degree of electrical conductivity are used on porcelain insulators to even out voltage differences and reduce interference effects. On certain types of bodies, glazes with very low thermal expansions are needed.

## 13.5 PROPERTIES AND COMPOSITION

The relationship between composition and physical properties in glazes is extremely complex. If we consider, for example, the case of coloured glazes, we know that characteristic colouring effects are associated with the selective absorption or emission of certain wavelengths of light by the colouring ions. These wavelengths are a function of the energy levels which are a characteristic of the particular ion. But this ion cannot be considered in isolation. In the glaze it is surrounded by a network of other ions, which modify the field, and thus the energy levels involved. For this reason, the same ion in different environments will give rise to different colours. Theoretical treatment is thus extremely complex, though some progress has been made in this field in recent years.

The design of glaze types is thus based on long experience and much empirical data. In this way the function of different oxides has become well established. It is thus well known that most glazes will need to have a high silica content to form the network. This network can be modified by such ions as sodium and potassium, thus substantially lowering the melting point, but having the disadvantage that they have a high thermal expansion. Lead, with a lower thermal expansion and a high refractive index, is very effective in producing good glaze flow, smooth surfaces and high degree of gloss. For high temperature glazes, calcium oxide is a good flux. Boric oxide is often used as a flux; alumina increases viscosity and inhibits devitrification. Colouring oxides commonly used include iron, copper, cobalt, chromium, manganese, nickel, vanadium, cadmium, and selenium. Zinc oxide has a beneficial effect in many coloured glazes.

Many attempts have been made to relate physical properties to composition by the use of factors allocated to particular oxides. One of the simplest examples is that of thermal expansion. It is assumed that the thermal expansion can be calculated from a simple linear relationship using factors for each oxide. Factors that are commonly used for glass were published many years ago by English and Turner (1929), and if we apply these to the glaze shown in Table 13.3, we obtain the result shown in Table 13.4. The total expansion to 500 °C would be 0.37 per cent, which would be of the right order for this glaze.

This kind of calculation is very limited in its application. For example, the same oxide will have different effects in different environments, and in any case, thermal expansion is not a fixed parameter, but is dependent on temperature, among other things. Nevertheless, the analysis does highlight certain features,

**Table 13.4** – Calculated thermal expansion for a glaze

| Oxide | Weight proportion | English and Turner thermal expansion factors | Product |
|---|---|---|---|
| CaO | 0.066 | $163.0 \times 10^{-7}$ | $10.78 \times 10^{-7}$ |
| $K_2O$ | 0.028 | 390.0 | 10.92 |
| $Na_2O$ | 0.055 | 416.0 | 22.88 |
| PbO | 0.132 | 106.0 | 13.99 |
| $Al_2O_3$ | 0.105 | 166.7 | 17.50 |
| $B_2O_3$ | 0.082 | −65.3 | −5.35 |
| $SiO_2$ | 0.532 | 5.0 | 2.66 |
| | | Total | $73.38 \times 10^{-7}$ |

such as the importance of $K_2O$ and $Na_2O$, which can be a useful guide in glaze design.

It is of interest to note that the advent of the computer has enabled a much more sophisticated approach to be adopted. It is never possible to make a glaze in which all the relevant properties are favourable, so a compromise has to be reached. The use of more complicated regression formulae enables the choices to be analysed and optimized in such a way as to save a lot of experimental investigation. Astbury (1969) has examined the whole question of the relation between composition and properties for ceramic materials.

## 13.6 THE MATURING OF GLAZES

In firing a glazed article, except in the case of matt glazes, especially in the case of fine ceramics, the aim is to produce a clear glass layer free from defects and having a smooth surface giving an attractive glossy appearance. It is now known that poor appearance is associated with an unacceptable amount of bubble in the glaze. This bubble may be uniformly distributed throughout the thickness of the glaze, in which case the appearance will be generally dull. In other cases, the surface of the glaze may be cratered, where a bubble has burst and healed, or, more usually, indented where there is a bubble under the surface, giving rise to a surface often referred to as 'eggshell' or 'orange peel'.

Many experimental methods have been developed for studying the way these bubbles develop, and are eliminated, including filming of the maturing process, so that the life history of a glaze is now well understood. It is well established that the gas in the bubbles comes almost entirely from the interstices between the unfired particles, with evolved gases from constituent materials, or pores within the body, playing only a very minor role. The main features of the maturing process have been fully described by Franklin (1965, 1966) and by Lepie and Norton (1960). Some features are illustrated diagrammatically in

Fig. 13.2, which shows the bubble distribution through a section of the glaze.

At (a), we see the unfired glaze, with the interparticle spaces which are unavoidable when the glaze is applied in powder form. These spaces can often represent as much as 50 per cent of the volume. As soon as the fusible particles begin to melt, some of the spaces are sealed off in the form of irregular shaped pores, as shown in (b). When fusion is well under way, at about 750°C for a

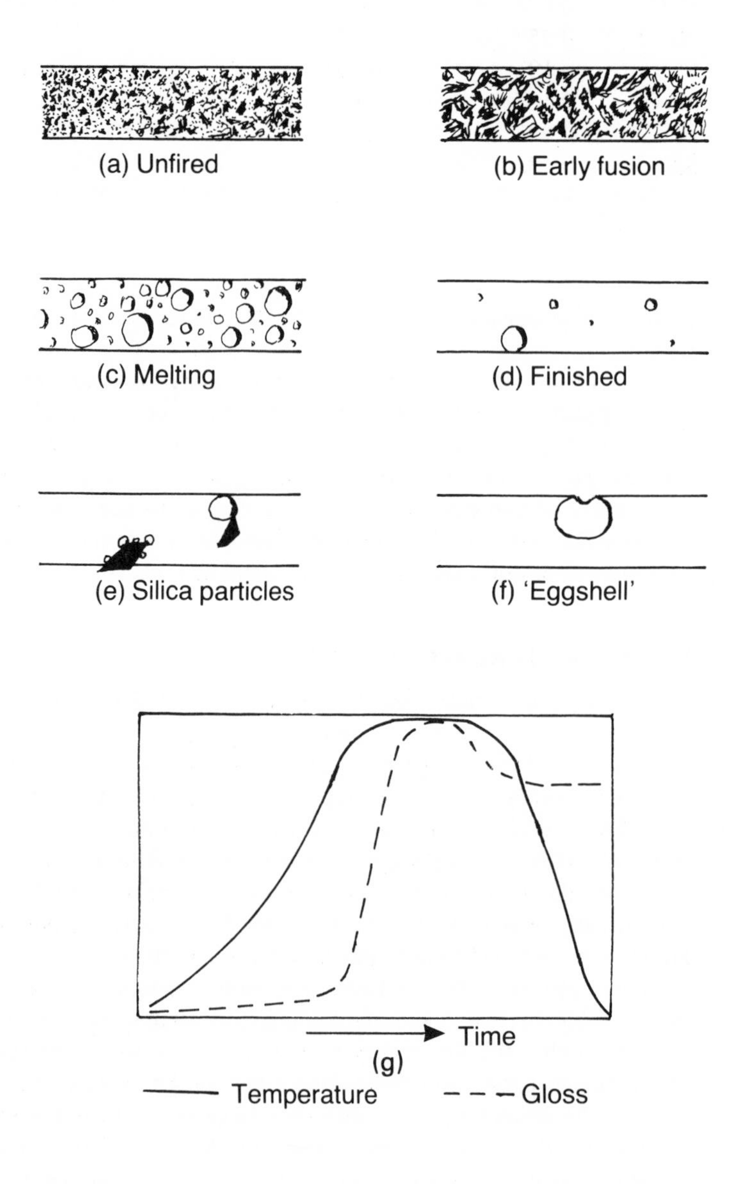


Fig. 13.2 – Maturing of a glaze.

low-sol glaze, surface tension forces turn these pores into spherical bubbles, which are very numerous, and cover a wide range of sizes, as shown in (c). From then on to the finishing temperature, the aim is to reduce the viscosity of the glaze so that the bubbles can escape. As either temperature or time increases, bigger bubbles are formed, though the total bubble volume does not change much. At the peak temperature, the viscosity is so low that the bubbles escape at the surface, and the surface heals very rapidly. Indeed, observation shows that the surface condition is better than it is on the finished article. There is always a slight deterioration in the early stages of cooling, though this can be minimized by slow cooling over the first 150° or so. The final condition is as shown in (d), with a few bubbles always remaining without noticeably harming the appearance. One of the interesting features emerging from these studies is that the underlying body has a marked effect on the appearance. The same glaze on a substrate of alumina, bone china, and earthenware, will give a residual bubble content increasing in that order, and a visual appearance declining in that order. The main reason for this is that bubbles attach themselves to large particles of silica, which may be in the glaze, or protruding from the body surface, and are very difficult to remove. There are none of these particles in the alumina, only a few in the bone china, and many in the earthenware. The effect appears to be chemical rather than physical. Alumina particles do not show it; silica particles do show it, whether they are crystalline or fused. Peripheral changes in viscosity due to changes in composition may be the key factor. A typical case is illustrated in (e). When a large bubble is retained near the surface, beyond the peak temperature, the gas volume is reduced as the temperature falls, giving the indented eggshell appearance, as shown in (f). This explanation is confirmed by the fact that if the glaze is cooled under reduced external pressure, the indentation does not occur. The overall change in appearance over the whole of the firing cycle is illustrated notionally in (g), which shows the optimum at the peak temperature.

## 13.7 GLAZE FIT

Glazed ceramic articles invariably require that the glaze on the finished product shall be in a state of compression. There are a number of reasons for this. Stresses imposed during the service life often put the surface in tension. Examples are the effect of size changes in the background to which wall tiles are fixed, and static or dynamic stresses caused by loading or impact. In the case of porous bodies, there is the possibility of expansion due to moisture over a period of time, thus tending to put the glaze in tension, resulting in crazing. Most ceramic materials are weaker in tension than in compression, so resistance to fracture can be improved by having a degree of compression in the glaze. In some cases, the strength can be increased by as much as 20 per cent by this means.

In order to achieve this compression, bodies and glazes have to be designed so that the thermal expansion of the body is higher than that of the glaze at its softening point. The softening point of the glaze is a rather loose concept. If we imagine the glaze cooling down in the kiln, there is no stress when it is in the liquid state. At a temperature of around 500-600 °C it stiffens fairly rapidly, acquires a degree of elasticity, and is capable of exerting and accepting stress. If the expansion of the body is higher than that of the glaze at that point, the body will tend to contract more on cooling to room temperature, and will thus put the glaze in compression when cold. A difference in expansion of the order of 0.1 per cent is usually adequate. For vitreous bodies, with no moisture

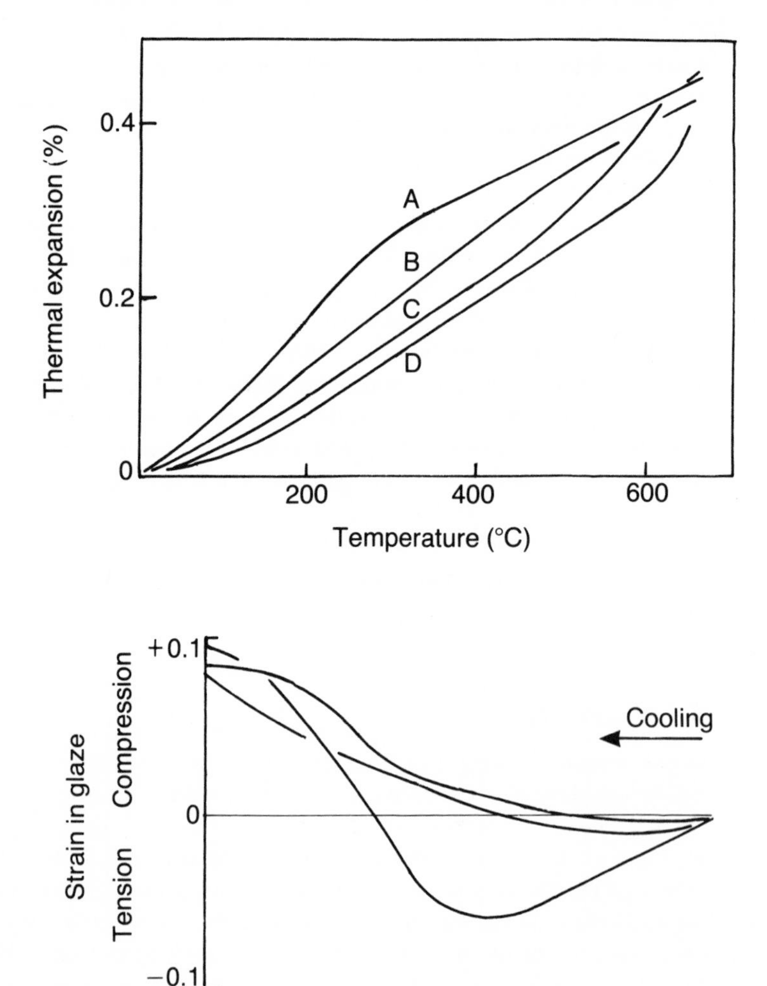


Fig. 13.3 – Development of glaze compression on cooling. A, Body high in cristobalite; B, bone china body; C, body high in quartz; D, glaze.

expansion, rather less than this is adequate. There is an upper limit to the thermal expansion difference that can be used in practice; if the compression is too high the glaze will peel away from the body.

The glaze compression does not develop immediately, but progressively over the cooling range according to the type of body. Fig. 13.3 shows the notional thermal expansion curves for three bodies, and a glaze with a softening point at 600 °C. Body A has a high cristobalite content; consequently, the compression does not develop until the contraction associated with the cristobalite conversion takes place around 250 °C. Indeed, prior to that there is an appreciable amount of tension. In the case of the high quartz body, C, the quartz inversion at around 570 °C means that the compression begins to develop in the early stages of cooling. The bone china body has no marked inversions, so there is a steady increase in compression during most of the cooling period.

The question of glaze fit is important for all glazed ceramics, but presents a formidable problem in the field of low thermal expansion bodies. These bodies have many potential advantages, for example where good thermal shock resistance is required, as in cooking. There is no real difficulty in disigning bodies with very low thermal expansion, but it is difficult to make glazes with a thermal expansion lower than 0.25 at 500 °C. These considerations similarly inhibit the use in bodies of a number of quartz free minerals.

In the above treatment, it has been tacitly assumed that the compression in the glaze is uniform throughout its thickness. Simple mechanical considerations will show that this is not so, and that the strain varies throughout the thickness of both the body and the glaze. The distribution of strain is readily calculable,

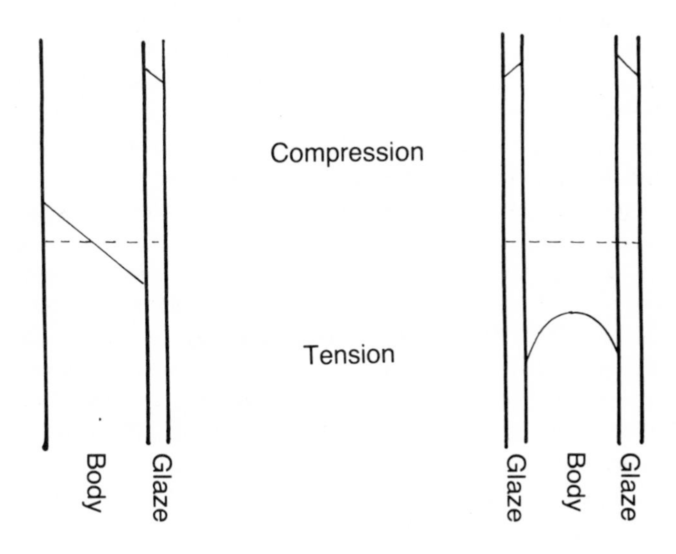


Fig. 13.4 – Variation of strain through glaze-body sections.

given the elastic constants and the thicknesses of body and glaze, together with the virtual strain generated by the thermal expansion difference (Hoens and Kreuter, 1965). Without going into the mathematical detail, we may note the general result for two simple practical cases shown in Fig. 13.4. The case of a body glazed on one side only is represented in practice by a wall tile. In this case, the combination acts in the same manner as a bimetallic strip, and the tile develops a convex curvature on the glaze. This change in curvature is allowed for in the spring of the die when the tile is pressed. The glaze compression increases towards the interface, where the body is in tension. The tension in the body decreases away from the interface; there is a neutral surface, and the surface of the body is in slight compression. In the second case, represented approximately in practice by a plate, there is glaze on both sides of the body, so there is no change in curvature. The whole of the body is in tension, with a maximum value at the interfaces.

Strain at the glaze surface can be measured directly by the use of strain gauges, or can be deduced from changes in curvature when the glaze is removed. (Vaughan and Dinsdale, 1960; Harrison, 1978).

## 13.8 BODY–GLAZE REACTION

We have assumed so far that the body and glaze constitute two distinct layers, each of uniform composition throughout the section, but in practice this is often far removed from the case. Body and glaze react with each other to form an intermediate layer, the thickness of which varies with the viscosity and firing temperature. In cases where the resulting mixture composition is appropriate, crystallization may take place; for example, mullite may often be found. In the case of bone china, the phosphoric acid in the body can react with the calcium oxide in the glaze to form apatite. The properties of this intermediate layer can be very different from those of either the body or the glaze; if the thermal expansion is different it may produce marked changes in strength. In addition to the formation of an intermediate layer, there are often marked changes in composition, and hence in properties such as thermal expansion and refractive index, throughout the glaze thickness.

In the early days, these changes were studied by optical examination of transverse thin sections, or by successively grinding off thin layers and measuring the properties (Smith, 1954; Norris, 1956). In recent years more definitive techniques have enabled the mechanisms to be better understood. EPMA studies have enabled element profiles to be established, and have confirmed the mobility of some of the ions in both body and glaze. Electrochemical measurements also indicate the kinds of reaction taking place (Roberts and Marshall, 1970). Some examples of glaze-body reaction, and variations in properties through the glaze, are shown in Fig. 13.5.

In (a), the variation in refractive index through a tile glaze is shown. A close empirical relationship between refractive index and thermal expansion can be shown to exist in any given system, and hence the changes in thermal expansion

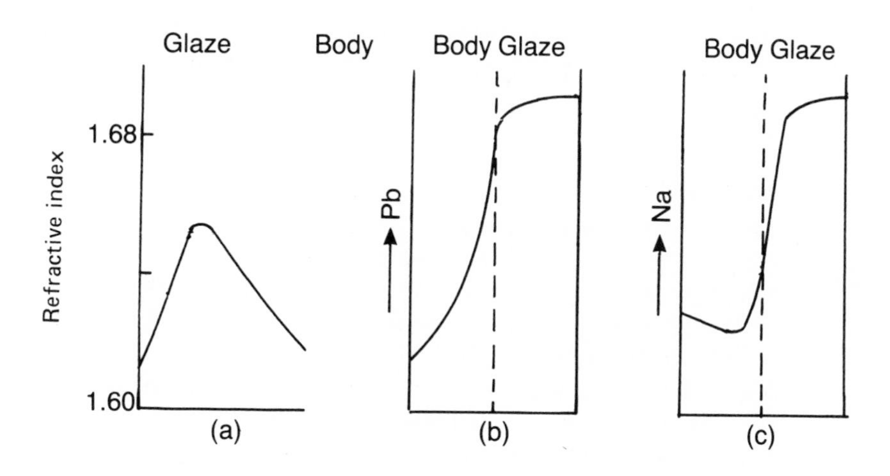


Fig. 13.5 – Variation through a glaze.

through the glaze can be inferred. In (b) and (c), the EPMA element profiles are shown through a glaze and into the bone china body. In (b) there is considerable penetration into the body by Pb from the glaze, stretching as far as 100 $\mu$m from the interface. Diffusion of sodium ions from the glaze into the body is illustrated in (c); it is also found that there is a corresponding transfer of potassium from the body into the glaze.

It is clearly established that considerable ion exchange takes place between body and glaze, and the concept of separate homogeneous layers is not strictly correct.

## 13.9 CHANGES AT THE GLAZE SURFACE

In addition to reactions at the interface, glazes can undergo composition changes at the surface, some of which can have a material effect on appearance. One cause of these changes is volatilization of glaze constituents; among the more volatile ones are Pb, Na, and B. In the course of a glost firing up to 10 per cent of the volatile components may be lost, with a result that the surface may have a dull or starved appearance. It is to prevent this that kiln furniture and saggars are often washed with glaze; this creates a saturated atmosphere, thus reducing the rate of loss. Another cause is the presence of gases or vapours in the kiln atmosphere that react with the glaze surface. When sulphur-containing fuels are used, the sulphates in the gases can be taken into solution by the glaze. If saturation is reached, crystallization will take place during the cooling, giving the well-known 'sulphuring' or 'feathering' fault. Different gases in the atmosphere affect the gloss in different ways; oxygen and water vapour adversely, and nitrogen and carbon dioxide beneficially. In some cases, the atmosphere is deliberately made reducing in order to produce a particular colour effect; copper reds are a well known example.

## REFERENCES

Astbury, N. F. (1969) *J. Brit. Cer. Soc.,* **VI**, 81.
Budworth, D. W. (1971) *tr. Brit. Cer. Soc.,* **70**, 57.
English, S. and Turner, W. E. S. (1929) *J. Amer. Cer. Soc.,* **12**, 760.
Franklin, C. E. L. (1965, 1966) *Tr. Brit. Cer. Soc.,* **64**, 549; **65**, 277.
Harrison, R. (1978) *Tr. & J. Brit. Cer. Soc.,* **77**(1), xii.
Hoens, M. F. A. and Kreuter, J. C. (1965) *Sci. of Ceramics,* **II**, 383.
Lepie, M. P. and Norton, F. H. (1960) *I.C.C.,* **VII**, 21.
Norris, A. W. (1956) *Tr. Brit. Cer. Soc.,* **55**, 674.
Roberts, W. and Marshall, K. (1970) *Tr. Brit. Cer. Soc.* **69**, 221.
Smith, A. N. (1954) *Tr. Brit. Cer. Soc.,* **53**, 219.
Vaughan, F. and Dinsdale, A. (1960) *I.C.C.,* **VII**, 89.

# 14

# Decoration

The word whiteware, often used to denote fine ceramics, is misleading. The intention is to distinguish between technical ceramics, refractories and structural clay products on the one hand, and naturally coloured single clay pottery bodies on the other, and the range of essentially white compounded body products with which we are principally concerned. In point of fact, by far the greatest proportion of these are not white in their finished appearance. In addition to fulfilling certain functional requirements, they rely for their customer appeal on aesthetic effect, resulting from attractive design and the artistic use of colour. This is especially true of tableware and ornamental ware, though it is also important in the case of wall tiles and sanitary ware. As with most consumer goods, there are fashions in colour and shape, changing through fairly short time cycles, so that the search for new decorative effects never ends. The decoration process is an important element in pottery manufacture, and can account for as much as 20 per cent of the production cost.

Decorative effect can be achieved in many ways. Bodies may be coloured; basalt and jasper ware are prime examples of this method, in which relief moulding can also play an important part. Glazes may be coloured, and much sanitary ware makes use of this technique. Mixtures of glazes can produce attractive random effects in tiles. On tableware especially, but also on tiles and, to a lesser degree, sanitary ware, decorative patterns composed of geometrical, natural, floral, or other themes, can be used to great effect. The ceramic background is as desirable a medium for the artist as canvas, as the fine work produced by direct on-glaze painting testifies.

The use of colour on or in pottery has been practised for centuries, and at some stages reached very high levels of attainment. The rich blue glazes of early Persian or Chinese ware have never been excelled. That these effects were achieved without any of the benefits of modern science and technology make them all the more remarkable. In this chapter we seek to describe the various decorating techniques, and at the same time outline the present understanding of the relationship between colour and the structure of the colour-producing elements. Much still needs to be discovered in this field, but enough is now known to provide a scientific base on which technological know-how may safely rest.

## 14.1 COLOUR SPECIFICATION

White light is a form of electromagnetic radiation that produces a sensation of vision in the human eye. It is made up of a continuous spectrum of wavelengths, each with its associated colour sensation. Selected parts of this spectrum can be separated out, either by emission from a medium, or absorption by a medium, and produce a sensation of colour in the eye. When the long wavelengths predominate, the colours are called red; at the short end of the wavelength range, they are called violet; in between, the colours range through orange, yellow, green, and blue. Not all colours, however, have the purity of the spectral colours. Diluted by white light, for example, the red becomes pink. In fact, three parameters are needed for a complete specification of a colour;

(1) Hue – the dominant spectral wavelength,
(2) Saturation – the purity of the colour, measuring the degree to which it is not diluted by white.
(3) Brightness – the intensity of illumination represented by the colour.

There are many systems available for representing these characteristics graphically, one of the most fundamental being the C.I.E. system, which has international status. In this system the co-ordinates of the colour are denoted by three coefficients, $x$, $y$ and $z$. Since these always add up to 1, it is only necessary to specify $x$ and $y$, and these can thus be plotted as rectangular co-ordinates on a graph. Such a C.I.E. chart is shown in Fig. 14.1. The continuous line represents the 100 per cent pure spectral colours. The wavelengths are indicated by numbers, the unit being nm. At the centre of the diagram is the white point, $W$, with co-ordinates $0.33x$ and $0.33y$. Any colour in the body of the chart can be represented by its $x$, $y$ values, or by the dominant wavelength and the saturation. The dominant wavelength, or hue, is found by projecting a line through the point from the white point, and noting the wavelength at which that line intersects the spectral locus. The ratio of the distances from the white point to the point in question, over the distance from the white point to the appropriate point on the spectral locus, is a measure of the degree of purity or saturation. For example, the point P on the diagram represents a pink colour, which is really

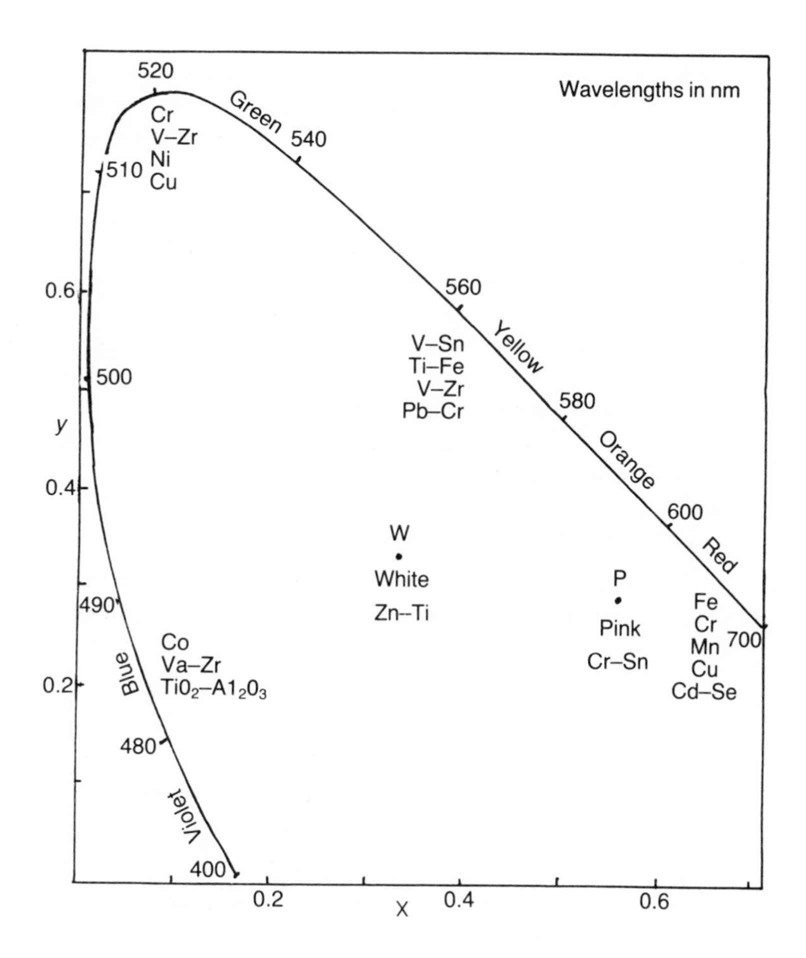


Fig. 14.1 – The C.I.E. colour chart, and colours associated with various elements.

a red diluted by white. It has a dominant hue of 600 nm, and a saturation of 60 per cent. Of course, this kind of chart tells us nothing about the brightness of the colour, which is the third important characteristic.

Also on the chart are shown some of the more important elements commonly used in glazes and ceramic colours and the areas in which the colours associated with them may be found. We need now to seek an understanding of the way in which the atomic structure of these elements determines the kind of colour produced.

## 14.2 THE ATOMIC BASIS OF COLOUR

The quantum theory model of the atom postulates that it consists of a nucleus surrounded by concentric shells of electrons. As the atom increases in atomic number, so the number of electrons increases, and they fit into a regular series of shells, each with its appropriate energy level. Four quantum numbers are used to express the electronic configuration of any particular atom.

(1) The principal quantum number, $n$, which may take the integral numbers 1, 2, 3, 4, . . . , proceeding outwards from the nucleus. These principal shells are often designated by the letters K, L, M, N, . . . .
(2) A secondary quantum number, $l$, representing the eccentricity of the electron orbit, and sometimes known as the orbital quantum number. The integral values of $l$ can range from 0 to $n - 1$. Letters often used to refer to these values are $s$ for $l = 0$, $p$ for $l = 1$, $d$ for $l = 2$, and $f$ for $l = 3$.
(3) For each of these values of $l$ there is a further subdivision relating to the orientation of the elliptical orbit, denoted by a quantum number $m$, which may take integral values from $-l$ to $+l$.
(4) Each energy level is further divided into two by the attribution of either a positive or negative value to the electron spin.

The Pauli exclusion principle states that no two electrons can have the same value for all four quantum numbers. Thus as the atomic number builds up, we have

| | |
|---|---|
| for $n = 1$ $l = 0$ only. | With two values for spin, this shell can only contain two electrons |
| for $n = 2$ $l = 0$ or 1 | For $l = 0, m = 0$, and there are two electrons. |
| | For $l = 1, m = -1, 0$, or $+1$, with two electrons in each |
| | There are thus eight electrons in the $n = 2$ shell |

Similarly, it will be seen that the $n = 3$ shell can contain up to 18 electrons. The nomenclature often used refers to the electronic configuration in terms of these numbers. Thus, an atom of nitrogen designated as $1s^2.2s^2.2p^3$. would have two electrons in the $1s$ shell, two in the $2s$ shell, and three in the $2p$ shell. A study of the periodic table shows that the chemical properties of atoms, and particularly the valency, are strongly related to the number of 'free' electrons above a completed shell. As the complexity of the atom increases, so does the energy level of the outer electrons. It is the transfer of electrons from one level to another, either emitting or absorbing energy, that gives rise to the phenomenon of colour.

## 14.3 COLOUR AND ENERGY LEVELS

When an electron moves from one energy state to another, the radiation that is emitted (or absorbed) has a frequency which is characteristic of that particular transition. If the quantum of energy is $\Delta E$, the frequency $\nu$ is given by the equation $\Delta E = h\nu$, where $h$ is Planck's constant. The frequency is related to the wavelength, $\lambda$, by the equation $\nu\lambda = c$, where $c$ is the velocity of light. From these two relationships it follows that the characteristic wavelength of the emitted radiation is related to the energy by the equation $\lambda = ch/\Delta E$, $\Delta E$ is usually expressed in electron volts, eV; one electron volt is the energy required to move an electron through an electric field produced by a potential difference of 1 volt. If we take the following values: $h = 6.6256 \times 10^{-34}$ Js, $C = 2.9979 \times 10^{8}$ $\text{ms}^{-1}$, and 1 eV = $1.6021 \times 10^{-19}$ J $\text{eV}^{-1}$ we have

$$\lambda = \frac{1239.8}{\Delta E} \text{ nm}$$

As an example we may consider the case of a free sodium ion, in which there is a transition between the 3s and 3p orbits. The energy difference between these levels is about 2.1 eV, and thus the associated wavelength is 589.3 nm, the well known sodium yellow. In point of fact, the 3p level has two energies very close together, corresponding to the two possible values of the electron spin. Thus the sodium D line, under high spectroscopic resolution, is found to be a doublet, with wavelengths of 589.0 and 589.6 nm.

The above considerations refer to atoms or ions in the free state; in gaseous form they are separated by relatively wide distances, so that the theoretical energy levels apply in practice. In solids, on the other hand, the ions are very close together and influence each other. As a result of this there is a good deal of overlapping, and the radiation is more in the form of continuous coloured

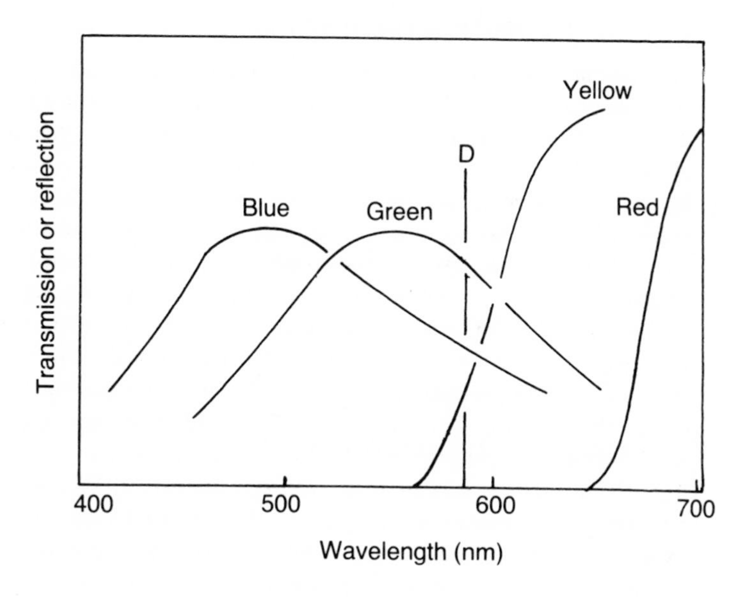


Fig. 14.2 – Transmission or reflection curves for various colours.

bands rather than in sharp lines. These continuous absorption or transmission bands do, however, have a dominant hue, and a colour characteristic of the element. This can be illustrated by using an alternative method of showing colour, in which the percentage of light transmitted or reflected by a material is plotted against the wavelength. When this is done it is found that many colours extend over large areas of the spectrum, as shown in Fig. 14.2, which also shows the sodium D line.

In physiological terms, colour sensation in the eye is dependent on three factors: the energy distribution in the light source, the absorption or reflection of the colouring element, and the sensitivity of the eye, at each wavelength. The total effect is obtained by integrating these elements across the whole visible spectrum. Looked at in this way, it becomes clear that a colour can be matched by two quite different reflection curves, but when the energy distribution in the light source is changed the match will not necessarily be maintained. This is the reason for the phenomenon known as metamerism, denoting the well known fact that colours matching in one light, such as daylight, look quite different under another source, such as electric light. Since different materials have different kinds of reflection curves, there is a problems in obtaining colour matches between, say, ceramic and plastic surfaces over a range of light sources. This fact makes it difficult to harmonize the colours where different types of material are in use, as in a bathroom, for example.

## 14.4 THE TRANSITION ELEMENTS

Most of the important elements used as a source of colour in glazes and enamel stains are to be found in a particular section of the Periodic Table known as the Transition Elements. The electronic configuration for the elements in this part of the table are set out in Table 14.1. In the s shells there are two available levels, corresponding to the two spins. In the p shells there are six available levels corresponding to $m = -1$, 0, and +1, with two spins for each. For the element Ar all these levels are occupied; there are no free electrons; so we have an inert gas. The 3d shell has 10 available levels, corresponding to $m = -2, -1$, 0, +1, and +2, with two spins each, and there are two levels in 4s. As we move to atomic number 19, we encounter the first anomaly. We might expect the extra electron for potassium to occupy the first 3d level, but in fact it is found in 4s. The reason is that the levels are now becoming very close together, with some overlapping, and the energy for 4s is slightly lower than that for 3d. The next electron also goes into 4s, so that 4s is full before there are any electrons in 3d. At atomic number 21, Sc, the 3d shell begins to fill, and we might now expect a regular sequence up to the 10 electrons required to fill this shell. However, another break in the sequence occurs at Cr, where two electrons go into 3d, one of them coming from 4s. At Mn, the 4s level is restored to its maximum, and the sequence continues until Cu is reached. At this point, there is a jump of two electrons in 3d, one of them again coming from 4s. At Zn, both the 3d and 4s shells are full. The elements Sc to Zn are known as the transition metals.

An important characteristic of these elements which is relevant to present considerations is that they all have multiple valencies. There are many possibilities of exchanges of electrons, and thus variation in colour. Moreover they readily form complexes, and the energy levels are distorted by surrounding ions. We shall examine this feature further in connection with coloured glazes.

**Table 14.1**

| | $n =$ | 1 | 2 | | 3 | | | 4 |
|---|---|---|---|---|---|---|---|---|
| | | K | L | | M | | | N |
| | $l =$ | 0 | 0 | 1 | 0 | 1 | 2 | 0 |
| Element | Atomic number | 1s | 2s | 2p | 3s | 3p | 3d | 4s |
| Ar | 18 | 2 | 2 | 6 | 2 | 6 | | |
| K | 19 | 2 | 2 | 6 | 2 | 6 | | 1 |
| Ca | 20 | 2 | 2 | 6 | 2 | 6 | | 2 |
| Sc | 21 | 2 | 2 | 6 | 2 | 6 | 1 | 2 |
| Ti | 22 | 2 | 2 | 6 | 2 | 6 | 2 | 2 |
| V | 23 | 2 | 2 | 6 | 2 | 6 | 3 | 2 |
| Cr | 24 | 2 | 2 | 6 | 2 | 6 | 5 | 1 |
| Mn | 25 | 2 | 2 | 6 | 2 | 6 | 5 | 2 |
| Fe | 26 | 2 | 2 | 6 | 2 | 6 | 6 | 2 |
| Co | 27 | 2 | 2 | 6 | 2 | 6 | 7 | 2 |
| Ni | 28 | 2 | 2 | 6 | 2 | 6 | 8 | 2 |
| Cu | 29 | 2 | 2 | 6 | 2 | 6 | 10 | 1 |
| Zn | 30 | 2 | 2 | 6 | 2 | 6 | 10 | 2 |

## 14.5 CERAMIC COLOURS

There are a number of colouring media used in ceramics, and they may differ from each other in fundamental ways. Four of the more important are considered here, in terms of the basic mechanism from which the colour effect derives.

It should be noted that the stability of colours varies greatly; in general terms, it decreases with increasing temperature. This means that at hard porcelain temperatures, around 1400 °C the range of available colours is much less than that obtaining with on-glaze enamels at 700 to 800 °C. Similarly, underglaze colours, which have to mature at the same temperature as the overlying glaze, have a restricted colour range. A relatively modern development is the in-glaze colour, which may be applied to either fired or unfired glaze surface, and which matures at glaze temperatures. Some elements have a high degree of stability; one such is cobalt, and this, together with its strong colouring effect, is the reason why it has such a wide range of uses in ceramic decoration.

### 14.5.1 Coloured glazes

Coloured glazes are essentially similar to transparent glazes, except that a small quantity, generally of the order of 5 to 10 per cent, and often much less, of colouring oxide is incorporated. The base glaze, however, has to meet a number of conditions. It has to be such that it matches the body for thermal expansion, but it has also to provide the correct environment for colour development.

It has already been noted that the main colouring elements, the transition metals, can readily form complexes in solution. This means that they can associate with two or more surrounding ions of opposite sign, sometimes known as ligands, to form a complex group, the properties of which may be very different from those of the original ion. The surrounding field can change the characteristic energy levels in such a way as to bring about a complete change of colour.

A simple example, often quoted, is the case of hydrated cobalt. In solution the cobalt chloride, $CoCl_2.6H_2O$, has a complex $Co(H_2O)_6$, in which six water molecules surround the central divalent ion in an octahedral arrangement, giving a characteristic pink colour. Treated with HCl, the anion $CoCl_4$, in tetrahedral arrangement, is formed and the change in environment changes the cobalt energy levels; and the colour changes to blue.

Study of the ligand field theory is held by many to be the key to understanding colours in glass and glazes, but this is by no means universally accepted (Taylor, 1967). That the nature of the base glaze can have a profound influence on the resulting colour derived from a particular element is a well-established fact. Thus copper in an alkaline glaze gives a turquoise colour, but green under acidic conditions. In entering the glass structure the colouring ion, such as Fe or Co, may take the place of $Si^{4+}$ in the network-forming function.

Some elements, such as Pb and Zn, do not themselves contribute any colour, but exercise a considerable influence on the colours produced by other ions. Some elements, like Cu are highly soluble in the glaze; on the other hand, $Cr_2O_3$, which gives a bright green colour, remains largely undissolved.

### 14.5.2 Stains

In contrast to the so-called solution colours a wide range of colouring effects can be produced by making use of crystals coloured by the transition ions, but insoluble in the surrounding glaze or flux. These are the stains that form the basis of most underglaze and on-glaze decoration. The stains themselves are prepared by an elaborate sequence of processes. The constituents are mixed wet, dried out, crushed, and calcined at a high temperature. Several wet and dry grinding processes may follow before the stain is mixed with flux. Control of particles size is crucial; too small and solution in the flux may occur; too large and the colour effect is lost. The quantity of flux may be very small, of the order of 5 per cent, in the case of underglaze colours. For on-glaze colours, where a good degree of fusion with the glaze is required, much more flux may be used – up to 60 or 80 per cent, according to the intensity of colour needed.

Most of these stains, though not all, are based on the use of spinel structures, which can give rise to a wide range of colours highly stable in the 750–850 °C range. Sometimes crystals have to be doped to produce a particular colour effect. The spinel structure itself has the general formula $A B_2 O_4$, where A is a divalent and B a trivalent cation. The unit cell contains 32 oxygen atoms; between these oxygens there are divalent cations in fourfold co-ordination and 16 trivalent cations in sixfold co-ordination. A modified version of this arrangement is known as an inverted spinel.

Some examples of spinels used in ceramic stains are listed in Table 14.2 together with their associated colours.

**Table 14.2** – Spinel colours in ceramic stains

| Spinel | Colour | Spinel | Colour |
|---|---|---|---|
| $CoAl_2O_4$ | Blue | $CaCr_2O_4$ | Green |
| $NiAl_2O_4$ | Blue | $CoCr_2O_4$ | Green |
| $FeAl_2O_4$ | Green | $MgCr_2O_4$ | Green |
| $CuAl_2O_4$ | Brown | $NiCr_2O_4$ | Green |
| $MnAl_2O_4$ | Cream | $ZnCr_2O_4$ | Green |
| | | $FeCr_2O_4$ | Brown |
| $ZnTiO_4$ | White | $FeV_2O_4$ | Black |
| $MnTiO_4$ | Brown | $MgV_2O_4$ | Black |
| | | $ZnV_2O_4$ | Black |

An important exception to the spinel structure is the use of Cd and Se as a basis for red and orange colours. In this case the colouring agent is a solid solution of CdS and CdSe. At higher temperatures $ZrO_2$ or $ZrSiO_4$ crystals can be used as hosts since they are not greatly attacked by the glass phase up to about 1250 °C. In conjunction with V they produce a blue colour; with Pr yellow; and with Fe pink.

### 14.5.3 Dispersed colours

Some ceramic colours depend on neither the solution of ions in the base flux, nor the development of spinel type structures, but are in the form of finely divided particles which attach themselves to a host or carrier crystal. A good example is the well known pink colour obtained when Cr is associated with $SnO_2$.

### 14.5.4 Metallic films

A widely used method of enhancing the attractive appearance of decorated ware is the use of precious metals, applied in the form of lines or bands, or even as screen-printed patterns. There are two main types of gold decoration. In so-called 'bright' gold, the gold is in the form of resinate, which breaks down during the firing, leaving a metallic layer of the order of 1-2 $\times$ $10^{-7}$ m thick. A much more durable form is the so-called 'burnished' gold, in which a much higher gold content is used, in conjunction with a flux, and the applied layer is much thicker, and hence more durable in use. This type of gold is dull after firing, and needs to be burnished by a hard stone or brush.

Metallic effects can also be produced by the use of 'silver' decoration, which contains mainly platinum or palladium, or a mixture of the two.

## 14.6 APPLICATION

The basis of all decoration is some kind of artistic theme, which may arise from any one of many possible sources; a picture, a natural scene, a pattern in the designer's mind, and so on. The problem of application is how to transfer this idea in the form of ceramic colours on to the surface of a piece of ceramic ware. The further problem, of great commercial significance, is that of repeating this transfer process accurately and effectively over a large number of occasions. The requirement is, of course, not dissimilar to that facing the printing industry, and it is thus not surprising to find that some of the techniques used in printing have found their way into ceramic decorating. But there are other techniques which are specific to ceramics. The range of methods used is very wide, and it is only possible here to deal with a limited selection in any detail. In describing these methods, a distinction is often made between under-glaze and on-glaze decoration, but, since many of the techniques are applicable to both, we do not divide them in this way, except to note where the differences are important. Four main groups may be distinguished, although many decorations require a mixture of two or more of them.

### 14.6.1 Surface treatment

Some types of decoration take the form of modifying the surface relief of the ware, in either a positive or negative sense. In some cases, the pattern may be cut out of the unfired body, or the same effect may be produced by pressing an indented pattern on tiles, for example. The indentations are then filled in with either a coloured body or a coloured glaze. This is the basis of the encaustic tile.

Conversely, a positive relief pattern may be included in the surface at forming, or subsequently added. The production of cameos is a case in point. When the moulding is added to a plane surface, it is produced from plastic body in a small mould, dried sufficiently to allow mould release, and stuck on the ware by means of a body slip, or by a slight application of water.

### 14.6.2 Direct application

The most obvious form of direct application is hand painting. The colours are compounded in a medium suitable for brushing, and the artist may create

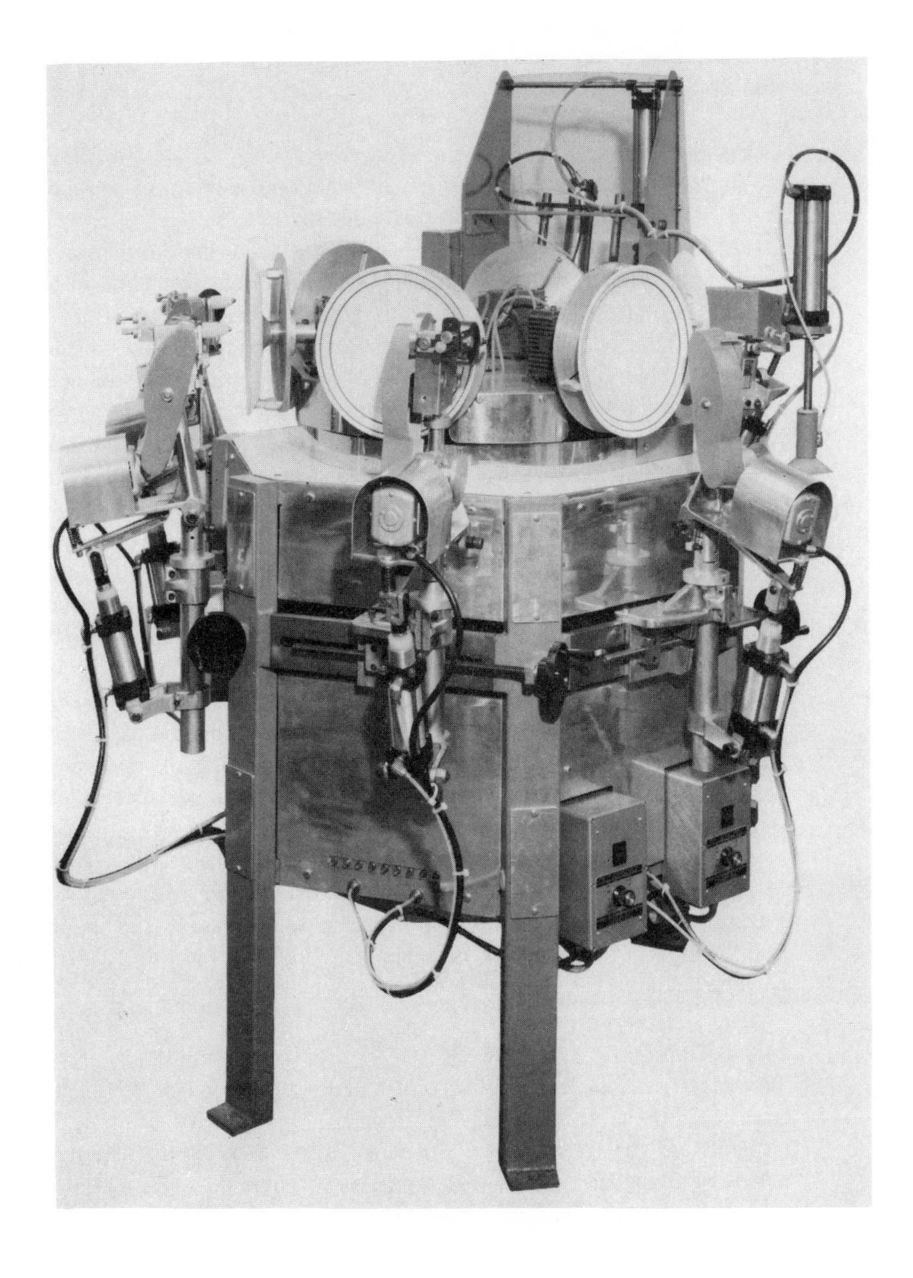

Fig. 14.3 – Lining and banding machine (Courtesy Malkin).

freehand painting on a one-off basis. This is highly skilled work, producing pieces of individual identity, and therefore only used in the case of expensive products. Hand painting is rather more widely used for in-fill work on printed designs.

Lining and banding, of gold or colours, can also be done by hand, assisted by a rotating turntable, but much of this work can now be done by machine (Fig. 14.3).

### 14.6.3 Stencils

A simple method of producing a decorative effect is to brush or spray a colour or glaze through a stencil of appropriate design.

A refinement of this technique, using the same basic principle, is the use of the 'silk' screen, which has made a very important contribution to modern pottery decoration. In this method, a tightly stretched screen, with a rectangular grid, of around 150 apertures per linear inch, is treated with a substance which blocks out the holes. The screen may be made of either nylon or polyester, or a stainless steel if higher temperatures are needed. The blocking out material is removed in selected areas corresponding to the pattern to be printed. A suitable modern method is to use a photosensitive gelatine, which can be exposed to a picture, so that the unexposed part can be dissolved out.

The screen is then placed near to, but not touching, the surface to be decorated, and the colour, in the form of a viscous ink, is forced through the screen by a squeegee blade. Successive layers of colour can be put on if they are allowed to dry between applications. The method can be used for either direct or indirect application. The direct application can be readily mechanized, as in the screen application to wall tiles. In the indirect technique, slide-off transfers are the most important use. One advantage that the method has, in comparison with other printing techniques, is that it enables thick layers of colour to be applied, and thus produces a characteristically 'solid' appearance. Improvements can be made by using thermoplastic inks, which are applied at around 70 °C, the screen being electrically heated using the metal mesh as a heating element.

Another form of stencil technique is that known as ground laying. The stencil is used to block out appropriate areas on the ware by the use of a soluble film. A layer of adhesive is then applied, followed by a dusting of the dry, powdered colour, the excess colour being blown off.

### 14.6.4 Transfer

The transfer principle is the basis of a range of decorating techniques, and is essential for the commercial development of repetitive printing. The main elements are the creation of a permanent plate carrying the design, a suitable means of filling the pattern with a colour-carrying ink, and a method of transferring the ink to the surface of the ware.

#### 14.6.4.1 *The plate*

The old method of producing the master plate was by engraving. For this purpose a copper plate may be used, with a hard chrome coating to reduce wear in use. The surface of the plate is cut by a variety of tools, producing lines or dots, so that the channels so produced can be filled with ink. The engraver can produce many tone effects by varying the frequency and size of the dots and the depth of the incisions. Engraving is a highly skilled operation, and the result has a distinctive character.

This technique is being replaced by the use of photographic methods (Basnett, 1980). A metal plate is coated with layer of photosensitive polymer; when this is exposed to a pattern of light and the soluble portion removed, the equivalent of an engraved plate is produced. The process is clearly much more rapid and more flexible than engraving. If half-tone effects are needed, they can be achieved by the use of interposed screens when the exposure is made.

14.6.4.2 *Lithography*

The lithograph process is now largely obsolete, but is described here for the record. It involves the printing of the colour on a paper backing subsequently transferred to the ware. The various layers involved are shown in section Fig. 14.4(a). The heavy backing paper consists of a thick outer

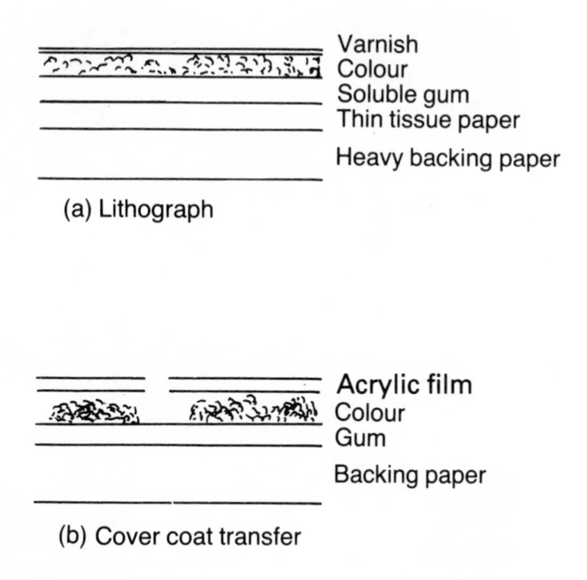


Fig. 14.4 – Methods of pattern transfer (a) lithograph, (b) covercoat.

Fig. 14.5 – Application of cover-coat transfer. (Courtesy Matthey Printed Products).

Fig. 14.5 – *continued*

layer combined with a thin tissue paper. On the tissue paper is a layer of soluble gum, on which the colour is deposited. This may be done either by direct printing, or by printing the pattern in the form of an adhesive and dusting on the colour. The layer of colour is then protected by a varnish. A lithograph pattern may be produced in the form of a large single sheet, on which are printed all the elements of the design that may be needed for a whole tea-set or dinner service. This means that the sheet has to be cut up into separate pieces, which have to be fitted to the ware. Since ware is not always uniform in size, skill is required to fit the pattern without obvious joins or other discontinuities.

In applying the lithograph, the backing paper is first peeled off, and the pattern is then applied to the ware which has previously been coated with size. In order to obtain good, uniform adhesion, the pattern has to be rubbed down hard with a suitable brush or other device, after which the ware is immersed in water so as to remove the soluble gum and the tissue paper. The varnish which is between the colour and the ware has to be burnt off during the firing.

14.6.4.3 *Cover-coat transfers*

Many of the disadvantages of lithography were overcome by the introduction of the slide-off transfer, leading to the widely used cover-coat system (Box, 1958). The arrangement is shown in Fig. 14.4(b). The colour is printed on to the soluble gum attached to the backing paper, either by lithograph or silk screen, and the acrylic film is set on top of the colour, not as a continuous layer but only in small areas sufficient to cover the motif. The transfer is slid off the backing sheet by soaking in water, and adjusted on the ware as shown in Fig. 14.5. In this form it is easily placed on curved surfaces, as well as flat.

14.6.4.4 *Off-set printing*

It is now about thirty years since the introduction of off-set printing in the pottery industry in the form of the Murray-Curvex machine (Pearce, 1958). In the traditional printing process, ink was transferred from the engraving plate to paper, and thence to the ware, either biscuit or glost. Even when the printed sheets were produced by means of a roller, the process as a whole was slow and tedious. In the Murray system, the paper was dispensed with, and transfer was effected via a gelatine pad. A simplified representation of the basis of the method is shown in Fig. 14.6, and a machine for two-colour printing, with automatic loader, is illustrated in Fig. 14.7. The pattern, in the form of an engraved plate, which may be of the metal or photopolymer type, is mounted on a support and fixed to a table, which also carries a mount for the article, such as a plate, to be printed. The table is capable of lateral movement. At the other side of the machine is a device for inking the pattern, together with a doctor blade for scarping off the excess ink. Between the two, mounted in a vertical mode, is a convex-shaped gelatine pad. The sequence of events commences with the table moving to the right and the pattern being inked. Then the pattern is returned to the central position and the pad is brought down. The flexible gelatine pad, under pressure, expands to cover the surface of the plate and to

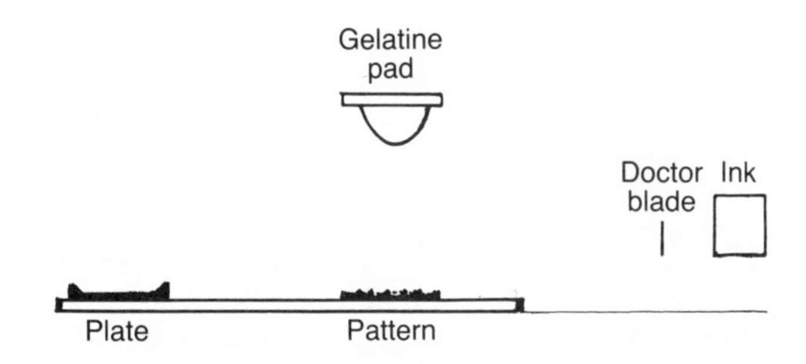


Fig. 14.6 – Off-set printing system.

take up the ink pattern. When the pad is withdrawn, the table is again moved to the right, so that the plate is aligned beneath the pad. The pad is then brought down and prints the ink pattern on to the surface of the plate. In recent years the British Ceramic Research Association has carried out a major successful research programme on all three elements of the system, plate, pad

Fig. 14.7 – Two colour off-set printer. (Courtesy Malkin).

and ink, paving the way for far-reaching developments in the industrial decorating process. The gelatine pad came to be used because it was cheap and easy to make and renew, but it had certain disadvantages that inhibited further development. These included poor mechanical strength, sensitivity to temperature changes, a tendency to pick up water, and poor elastic recovery after compression. The development of new silicone rubber pads has opened the way to improved performance and the possibility of automation.

The next problem was that the conventional inks were found to be incompatible with the silicone rubber, causing swelling in the area of contact, and consequent progressive loss of quality in the print. In the course of developing new inks, it was realized that the fact that the silicone rubber was stable to temperatures of around 100° C made it possible to use thermoplastic inks. These offered greater printing precision with little risk of contamination when overprinting colours in multicolour decorations. This in turn contributed to the concept of total colour transfer.

The third component in the system to receive attention was the plate. The old metal engraved plate was costly to produce, requiring a high degree of manual skill. Even acid-etched plates were expensive to produce, and limited in the amount of colour they could carry. The first major change was to replace the metal by a photopolymer plate, in which a photosensitive nylon, on a thin metal backing, is exposed to an image which produces polymerisation and hardening. Subsequent processing results in a plate that is relatively cheap, capable of carrying a wide range of designs, with a satisfactory durability in service. The facility for dealing with short-run customised designs is of great commercial significance.

In spite of the improvements brought about by the use of the photopolymer plate, there were still certain limitations. There is a limit to the weight of colour that can be transferred in any intaglio method, so a further advance depended on development in the plate area. The silk-screening of a pattern on to a flat surface seemed to offer most promise, but the greatest difficulty was that of ensuring a clean pick-up of ink from the plate, followed by total transfer from the pad to the ware. Problems of adhesion and viscosity were eventually solved by the use of thermoplastic inks, in which heated screens effected the required flow condition at screening, and the use of a silicone rubber substrate. If the substrate is maintained at a slightly lower temperature, possibly by water circulation, the cooler ink then develops the right degree of tack for total transfer to the pad, and then to the ware. With a different pad and substrate for each colour, multicolour designs can be built up. Since the ink hardens on the ware no cross-contamination occurs. A wider range of weight of ink deposit can be achieved, making possible the transfer of delicate half-tone blocks, lines, and continuous bands. The total colour transfer system is illustrated in Fig. 14.8.

These developments are rapidly revolutionising the decorating of pottery, and a further likely feature is the extension of printing on to clay ware. Lower pressures have to be used on account of the low strength of unfired body, and consideration has to be given to the use of special supports. But the commercial

Fig. 14.8 – Four colour total transfer printing machine (Courtesy Service Engineers).

incentive is such that developments in this field are likely to continue.

True trichromatic printing may well be a possibility for the future.

## 14.7 FIRING

The firing of underglaze colours is, of course, carried out during the normal glost fire; as is also the case with in-glaze colours. For on-glaze colours, however, a separate firing process is required. Indeed, if the palette consists of a wide range of colours maturing at different temperatures a series of firings, at diminishing peak temperatures may be needed. On hard porcelain the decorating fire may be up to 1000 °C, but on earthenware or bone china, the peak temperature is in the region of 750° C. A typical firing curve is shown in Fig. 14.9. It will be noted that the rate of rise of temperature is slowed down between 200 and 400 °C. This is in order to minimize the fault known as frizzling and to allow the organic matter associated with the decoration to be burnt off. Sufficient time is needed at the top temperatures to allow full development of the colour and a bright, glossy appearance. The cooling can be very rapid, consistent with the avoidance of dunting in bodies containing quartz or cristobalite. Attention has been given in recent years to the possibility of using much shorter firing times, 90 minutes having been shown to be possible. Bull (1982) has described the development of suitable colours for fast firing, using much higher peak temperatures.

One of the most persistent faults to develop in the decorating fire is that commonly known as spit-out. This takes the form of a large number of craters in the glaze surface resulting from the bursting of bubbles. The fault is particularly associated with porous bodies, such as earthenware. Wilkinson and Dinsdale (1961) have shown that the prime cause is the desorption of water from the internal surface of the biscuit, but two other essential ingredients are the presence of crystalline silica in the body, and the presence of bubbles in the glaze. Water vapour is taken in during glost storage, and the decoration process, and is continu-

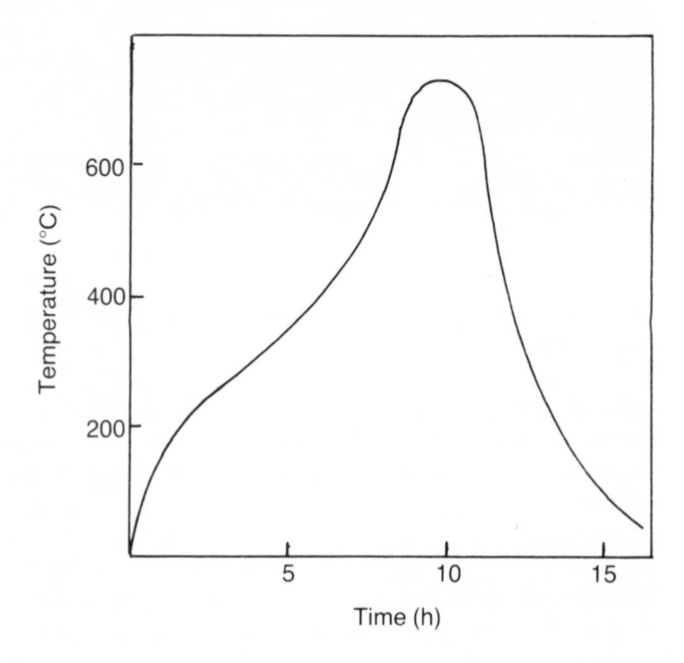


Fig. 14. 9 – Typical decorating kiln firing curve.

ously desorbed as the decorating kiln temperature increases. Pressure is thus built up, and easy access through cracks in the silica particles enables bubbles to expand and burst. As the temperature increases, the pressure increases and the viscosity of the glaze decreases, so that, unless time is allowed for the gas to escape by other means, there is an increasing tendency to spit. Thus, high enamel kiln temperatures and rapid rates of heating favour the development of the fault. For this reason, if for no other, the development of rapid firing for decoration is less likely to be successful with earthenware than with porcelain or bone china.

## 14.8 REACTIONS DURING FIRING

Since glazes and colours have much in common as regards composition, there is bound to be appreciable reaction between them during the decorating fire. To some extent this is desirable, as it is necessary to develop good adhesion between the two. On the other hand, running of colours has to be avoided, or the sharp delineation of pattern edges would be lost.

In fact, it is known that there is considerable migration of ions not only from the glaze, but also from the body, into the decoration. This is consistent with the marked effect that glaze composition has on the durability of the on-glaze decoration, and has been shown experimentally by the use of radioactive sodium to take place (Franklin *et al.*, 1960).

## 14.9 DURABILITY OF DECORATION

In service conditions, decorated tableware is faced with both mechanical and chemical attack, and durability is an important element in its acceptability to the customer.

Abrasion takes the form of contact with other objects, such as cutlery, but also contact between one piece and another. For example, when plates are stacked on top of each other, the foot of one plate rubs against the face of the other. It is very difficult to avoid this type of abrasion, since, however hard a glaze may be, it will scratch itself.

Chemical attack may be either by alkali or acid. The former has assumed importance in recent years owing to the development of domestic and industrial dish-washing machines. Although the attacking mechanism is very complex, general consideration of glass structure would predict that low temperature colours, having a more loosely bonded structure, would be more open to chemical attack. Not all colours are alike, however. Yellows and greens and some reds are less resistant to alkali attack than pinks; rich blues are well known to be vulnerable to acids. But within a given colour range there is often a choice of formulation, so that improved resistance can be maintained by careful selection. The most obvious means of avoiding the problem is by the use of under-glaze decoration or by using high temperature in-glaze colours. In both cases the colour palette available is more restricted than with low temperature patterns. In all cases, it is generally true that a good firing temperature, with the colours fully matured, gives the best durability.

Washing agents are easily classified into three types. For hand washing, normally carried out at temperatures no higher than 110 °F, there is no problem with modern household synthetic detergents. Soap powders or flakes, on the other hand, are corrosive. In domestic dishwashers, the powders are mainly compounded from sodium hydroxide, carbonates, silicates, or phosphates. Because of the need for air drying, these machines operate at temperatures of 140 °F or more, and attack becomes more likely at these higher temperatures. In a detailed study of this problem, Franklin *et al.* (1959, 1960) have shown that, whilst concentration of the washing agent around the recommended figure is not important, the attack increases rapidly with increased temperature of wash. The attack is not simply related to the alkalinity. The deterioration in colours is first noticeable in a loss of gloss, colour being bodily removed only in the later stages. Indeed, the measurement of the early loss of gloss is a fairly reliable index of the durability.

In industrial washing machines, the conditions are even more aggressive, since the washing has to be achieved in a much shorter time. The increasing use of on-glaze decorated ware in restaurants and institutions has brought this problem to the fore in recent years.

On-glaze colours are also attacked by acids, the most significant being organic acids associated with foodstuffs, such as vinegar or fruit juices. Oxalic acid is the most corrosive.

In general terms it has been shown that the underlying glaze can have a marked effect on the durability of the on-glaze colours. Thus, high silica or high lead in the glaze improve the acid resistance; high alkalis and alkali earths lower it. Alkalis, alkali earths, and boron tend to reduce the resistance to attack by washing agents.

## REFERENCES

Basnett, D. (1980) *Tr. and J. Brit. Cer. Soc.*, **79**(4), lvii.

Box, E. R. (1958) *Tr. Brit. Cer. Soc.*, **57**, 541.

Bull, A. C. (1982) *Tr. and J. Brit. Cer. Soc.*, **81**(3), 69.

Franklin, C. E. L. and Tindall, J. A. (1959) *Tr. Brit. Cer. Soc.*, **58**, 589.

Franklin, C. E. L., Tindall, J. A., and Dinsdale, A. (1960) *Tr. Brit. Cer. Soc.*, **59**, 401.

Pearce, N. (1958) *Tr. Brit. Cer. Soc.*, **57**, 527.

Taylor, J. R. (1967) *J. Brit. Cer. Soc.*, **4**(2), 201.

Wilkinson, W. T. and Dinsdale, A. (1961) *Tr. Brit. Cer. Soc.*, **60**, 33.

# 15

# Properties of finished ware — mechanical

The most important mechanical property of fired pottery is what may loosely be termed strength: that is, the ability to withstand without fracture the various stresses to which it may be subjected in use. Much research has been concentrated in recent years on the search for very high levels of strength in special ceramic materials for engineering use. We are concerned here, however, with somewhat lower strength levels, though many of the established principles still apply.

In the service life of pottery products, the compressive strength is of no significance; if articles fracture it is generally in tension. The stresses encountered may be either static or dynamic, as a few examples will illustrate. Tableware may be exposed to high static stresses when stacked; or may be subjected to impact stress when dropped on a hard floor. The impact is often of such a drastic nature that no conceivable practical strength increase would have prevented fracture. This kind of catastrophic damage is the principal mechanical hazard which tableware has to contend with.

At the other end of the spectrum, electrical porcelain components frequently have to sustain very heavy static loading; insulators on transmission lines are a case in point. These are generally designed to make use of the high compressive strength of procelain, but severe bending stresses are often encountered in posts or hollow support insulators. Dynamic stresses are met in many uses, such as switchgear.

For sanitary ware, heavy articles dropped into wash-basins represent a frequent hazard; overtight tap or location fittings can also give high localized stresses. Wall tiles have to contend with collisions from wheeled traffic, and may also be affected by large stresses imposed by the backing structure, which may undergo appreciable size changes over a period of time. Contraction in cement is a frequent source of trouble.

In most of these cases, the service life of the product may be significantly improved by increases in strength, and in this chapter we examine the physical factors that influence strength in whiteware bodies, and consider what practical steps can be taken to effect improvement.

## 15.1 DEFINITIONS AND MEASUREMENT

In this context we are mainly concerned with tensile strength, and the particular aspects of it that is involved in bending. This strength can be measured by subjecting a rod of material to tension, but the practical difficulties of this method are formidable. More usually, the rod is subjected to bending, and the stress to produce fracture is fairly easy to measure. Four-point loading is often preferred on the grounds that it involves a constant bending moment along the length of the rod. The strength thus measured is often called the cross-breaking strength or the modulus of rupture. Measurement of the surface strain will also enable a value of Young's modulus to be obtained.

The materials with which we are concerned are brittle: that is, there is elastic behaviour up to the point of rupture, with the stress–strain relationship being linear, in accordance with Hooke's law.

If we denote the applied stress by $\sigma$ and the resultant strain by $\epsilon$, we have $\sigma = E\epsilon$, where $E$ is Young's modulus. The unit of stress is $Nm^{-2}$, and since strain is dimensionless, the unit of Young's modulus is also $Nm^{-2}$. Typical values of $E$ for whiteware materials range from about $5 \times 10^{10}$ $Nm^{-2}$ for weaker porous bodies to about $10 \times 10^{10}$ $Nm^{-2}$ for bone china and porcelain.

## 15.2 THE STATISTICAL NATURE OF STRENGTH

Measurements on modulus of rupture on a sample of rods generally show considerable variation from piece to piece, with a standard deviation of perhaps ±10 per cent or more. Strength has thus to be understood in statistical terms. A determining factor is the presence of flaws in the material, which initiate cracks. If one such flaw occurs in the surface of a rod, on the side subject to maximum tension, a low value for the strength will result. The Weibull theory relates the probability of flaws being present within a given element of body to the stress operating within that volume, and predicts how the measured strength will depend on the flaw distribution, and the stress distribution in a particular method of measurement. Certain important practical conclusions follow. One is that the measured strength of a particular material is found to vary with the size and shape of the specimen and the method of applying the stress. In particular, tensile tests tend to give lower values since there is a large volume under maximum stress. Other predictions confirmed by experiment are that the mean strength decreases as the sample size is increased, and that the variation increases as the mean strength increases. It is therefore important when quoting strength values that the experimental conditions are clearly stated.

## 15.3 CRITICAL STRAIN

If the values for Young's modulus and strength are plotted for a wide range of ceramic materials it is found that there is an approximately linear relationship between the two quantities. The physical meaning of this is that over a wide range of strengths fracture occurs at a fairly constant value of the strain. The value of this critical strain is of the order of $10^{-3}$. It is thus possible to predict the strength of a material with reasonable accuracy from the value of Young's modulus. Thus, for example, porcelain, with $E = 10 \times 10^{10}$ Nm$^{-2}$, the strength would be $10 \times 10^{10} \times 10^{-3} = 10^8$ Nm$^{-2}$. An alumina body, with $E = 35 \times 10^{10}$, would have a strength of about $3.5 \times 10^8$ Nm$^{-2}$.

## 15.4 LEVELS OF STRENGTH

It is possible to divide the idea of strength in ceramic materials into three levels.

At the highest level there is the theoretical strength. This is the strength that would be obtained with a perfect crystalline material, free from flaws or imperfections of any kind. In this concept, the force required to cause fracture is that which is required to separate adjacent planes of atoms, and to supply the energy for the creation of new surfaces. Estimates of this theoretical strength generally suggest that it is about $7 \times 10^9$ Nm$^{-2}$. This strength level is not realizable in practice: the next level down is technically possible, with strengths of the order of $7 \times 10^8$ Nm$^{-2}$. This level can be reached with pure polycrystalline materials, sintered to maximum density and free from flaws. Further down the scale by another factor of 10, we come to industrial whiteware ceramics. Because of the presence of pores and flaws the strength of these materials is well below the technically possible level, and is usually in the range 5-10 $\times$ $10^7$ Nm$^{-2}$.

This level of strength is generally adequate for the use to which the products are to be put, but any increase in strength is welcome, providing it can be achieved within the terms of normal factory practice. Improvement has to begin with the recognition that the strength of these materials depends on texture rather than structure. The body consists of a glassy matrix in which are embedded crystals that may only approximately match the matrix in expansion. In addition to the resultant microcracks there are also pores, even in what are known as vitreous bodies. The flaws and pores are the principal factors in the reduction of the strength below the technically possible. We consider now some of the ways in which strength can be increased.

## 15.5 POROSITY

The firing of whiteware bodies results in a reduction in porosity as shown in

Fig. 11.2. In the case of earthenware, the fired porosity may be of the order of 15 per cent. On the other hand, in the case of vitreous bodies the apparent porosity may be reduced to less than 1 per cent. However, there is always a residual true porosity, in the form of closed pores, of about 5 per cent. The fact that these pores are closed does not diminish the effect they have on strength.

The presence of pores in the body always results in a reduction in strength. A number of formulae have been suggested for the relationship between strength and porosity, but experimental results usually indicate that an exponential expression fits the facts with sufficient accuracy. Thus, if $S$ is the strength at a fractional true porosity $p$, and $S_0$ is the extrapolated strength at $p = 0$, we can write $S = S_0 \, e^{-kp}$, or $\log S = \log S_0 - kp$, where $k$ is a constant. In many practical cases a plot of $\log S$ against $p$ gives a good straight line.

Dinsdale and Wilkinson (1966) have examined the results for a large number of earthenware and china bodies. They found that, although these bodies had different recipes, and were fired under different factory conditions, the results fell on remarkably consistent straight lines. The derived equations from these results were

for earthenware $S = 1.221 \, e^{-3.90p} \times 10^8 \, \mathrm{Nm^{-2}}$ (15.1)
and for china $S = 1.255 \, e^{-3.63p} \times 10^8 \, \mathrm{Nm^{-2}}$ (15.2)

over the range $p$ = 0.05 to 0.35 for china and 0.10 to 0.33 for earthenware. These two lines are shown in Fig. 15.1. It is remarkable that although these two body systems are very different chemically, and in other ways, they show almost identical strength-porosity relationships. Thus the fact that bone china

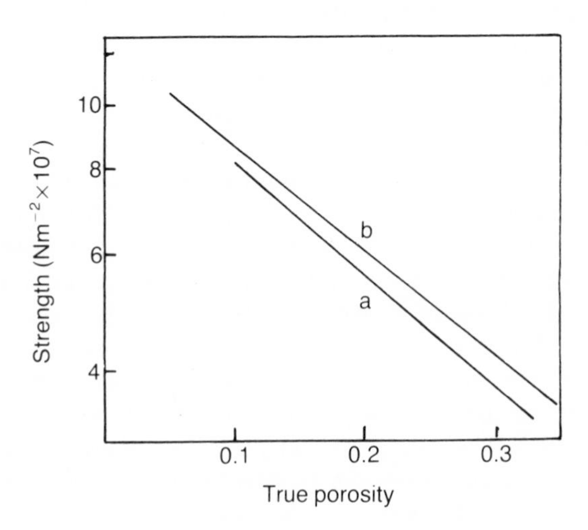


Fig. 15.1 – Relationship between strength and porosity for (a) earthenware and (b) china bodies.

as produced is much stronger than earthenware owes a great deal to its lower fired porosity.

The last 5 per cent of true porosity is difficult to eliminate. Under special atmospheric conditions it can be reduced, but in practical industrial conditions it has to be tolerated. It is perhaps not generally realized that, because of the logarithmic law, it means a loss of potential strength of about 15 per cent.

## 15.6 GRAIN SIZE

Present-day understanding of fracture mechanisms in brittle solids has been built on a development of ideas put forward more than half a century ago by Griffith. The problem is to explain why the presence of flaws results in such a large reduction in strength. Two considerations are involved. Cracks can act as stress-raising agents, so that at the tip of a crack the stress can be very much higher than the general stress distribution throughout the body of the material as a whole. Then, once a crack has been initiated, it can continue to grow only if the available elastic strain energy in the volume of material adjacent to the crack is sufficient to provide the extra surface energy associated with the newly fractured surfaces. Both these considerations lead to the conclusion that the stress required to produce fracture will vary inversely with the square root of the crack dimension. Thus if $S$ is the fracture strength and $c$ the crack dimension, $S \propto c^{-\frac{1}{2}}$.

In the kind of whiteware systems we are considering the initiating crack may be in the surface of the material, but it may equally well be inside the body. It is known to be associated with the filler component, and may derive from any one of a number of circumstances. Thermal mismatch between the particle and the matrix may cause cracks either in the filler or in the interface. Crystalline inversion in quartz particles may be the cause. Further, it is not always realized that all ground materials must contain cracked particles, since the grinding stops at an arbitrary point at which further size diminution is still taking place.

In applying the Griffith reasoning we make the assumption that the crack dimension is related to size of the grain to which it corresponds. In that case, if we denote the grain size by $G$, we may expect to find $S \propto G^{-0.5}$. We note here that we have a power relationship, as distinct from the exponential relationship we found between strength and porosity. What evidence is there for practical systems that grain size has this kind of effect?

Comparative strength measurements on a range of earthenware bodies containing flint varying in mean size from 3 to 8 $\mu$m, and using the strength extrapolated to zero porosity as the criterion, gave a value of the power index of 0.4. Interesting results were found from a series of experiments using alumina as the filler. The various grades of alumina were prepared so that each had a narrow size range, the median size of the grades ranging from 50 $\mu$m down to 5 $\mu$m. The results are shown in Fig. 15.2. It will be seen that over a wide range of sizes there is a linear relationship between the strength and the inverse square root of the grain size. Batchelor and Dinsdale (1960) reported on a range of alumina bodies, with sizes varying from 4 to 15 μm diameter; analysis of these results yields a value for the index of 0.75. So these three sets of results on different bodies give values of the index of 0.4, 0.50 and 0.75, which are in reasonable agreement with the theoretical 0.5.

In many of these experiments it is often found that the above relationships between strength and porosity, and between strength and grain size, break down at very low values of the grain size and at higher temperature. Looking again at Fig. 15.2, it will be noted that the increase in strength due to decreasing grain size ceases at the finer end of the size range. This is a particular manifestation of a general observation that the simple exponential and power relationships do not hold when there is an appreciable amount of reaction between the matrix and the filler, which is likely to be the case when the filler size is very small or the firing temperature is very high. In these circumstances, when solution and recrystallization occur, the filler tends to lose its identity. One feature of this situation is that the maximum strength may not occur at zero porosity, but at a higher porosity in an over-fired condition. In spite of these reservations it

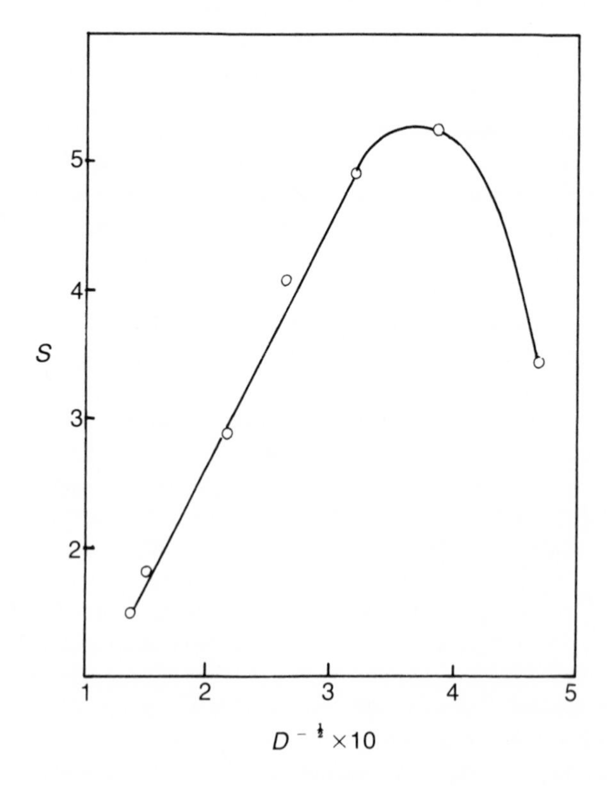


Fig. 15.2 – Relationship between strength, $S(\text{Nm}^{-2} \times 10^8)$, and grain size, $D(\mu\text{m})$, for alumina bodies.

remains true that, for normal grain sizes and normal firing, the strength increase resulting from decreased grain size is in line with the theoretical power law for a wide range of commercial bodies.

Examination of production bodies under the microscope shows the presence of a number of larger quartz particles, often up to 150 $\mu$m in size; that these result in a strength reduction can be demonstrated by deliberately adding them to a body in small percentages, resulting in substantial drop in strength. In considering the best method of removing them, it should be noted that most of them come from the quartz component of the body, but some may come from the flux, and a few perhaps from the clay. Large particles can be removed from the quartz either by fine grinding or by the use of a hydrocyclone, and this can produce strength increases of up to 100 per cent. Beech and Norris (1961) investigated the effect of fine grinding of the whole of the body, and confirmed increased strength. However, they also found that other properties were affected, as might be expected. The vitrification temperature was reduced, but a large increase in contraction was also encountered. The latter is a serious disincentive to this approach. On the other hand, a positive result is the improved surface on articles made from bodies from which large quartz particles have been removed.

## 15.7 FILLERS

### 15.7.1 Elastic modulus

One means of increasing strength that has been explored is the substitution of quartz by a filler of higher Young's modulus, such as alumina. In predicting the change in Young's modulus of the body, two models are commonly used. In one the matrix and the filler are assumed to carry equal stress; in the other they are assumed to share the strain equally. If a filler of Young's modulus $E_2$, is contained in a matrix of Young's modulus $E_1$, and the volume concentration of the filler is $c$, then the Young's modulus, $E$, of the composite material is given by

$$E = \frac{E_1}{1 - c\,(1 - E_1/E_2)}$$

for the constant stress model, or

$$E = cE_2 + (1 - c)\,E_1$$

for the constant strain model.

The constant strain model gives the higher value. Work by Binns (1962), using hot-pressed compacts of glass and alumina, showed that the experimental results fell between the two predicted values, with some bias towards the constant strain model.

It is of interest to consider a practical case, in which quartz is replaced by alumina, the concentration of filler in the body being $c = 0.35$. Approximate values of $E$, in $\mathrm{Nm}^{-2} \times 10^{10}$, can be taken as 4.8, 6.2 and 27 for the matrix, quartz, and alumina respectively. Substituting these values in the equations, there would be an increase in $E$ by a factor of 2.4 on the constant strain model, and 1.3 on the constant stress model. In fact, experiments show that the increase in strength is about 2:1. It needs to be remembered, however, that this assumes that the constant strain criterion applies, and no account is taken of strength improvements resulting from the elimination of quartz inversions. Nevertheless, alumina-containing bodies of high strength are commercially produced. That they are not more widely used in tableware may be due to the disadvantages of appreciably increased weight and cost. In technical porcelains, where strength is very important, these disadvantages can be more readily accepted.

### 15.7.2 Thermal expansion

Thermal mismatch between filler and matrix results in the presence of stresses that may produce cracking and a lowering of body strength. The general effect is as shown in Fig. 15.3 (Binns, 1962). Optimum strength is found when the expansions are equal, decreasing on either side but more seriously when the filler expansion is low. In this case the tendency is for radial cracks to form, whereas in the case of high expansion fillers the cracks are usually tangential. The thermal expansion of quartz is about twice that of alumina.

### 15.7.3 Fibres

Modern ceramic technology is making increasing use of the fact that fibres, or 'whiskers' as they are sometimes called, can show abnormally high strength. It is tempting to suppose that the incorporation of these into whiteware bodies would increase the strength. Naturally occurring minerals in the form of laths, with aspect ratios of around 10:1 have been tried for this purpose. The results have been disappointing, mainly due to the difficulty of arranging the right orientation. In order to obtain maximum effect the fibre needs to be at right angles to the direction of crack propagation. However, it so happens that, for

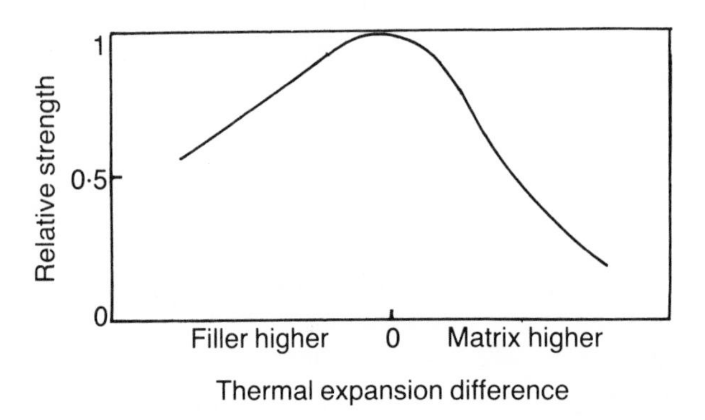


Fig. 15.3 – Effect on strength of thermal expansion of filler and matrix.

example, in the plastic shaping of plates the tendency is for the fibres to be orientated radially when, in order to improve resistance to chipping, they need to be parallel to the edge.

## 15.8 GLAZING

The principal reason for glazing whiteware bodies is to improve the finished appearance, but, in addition, the effect of glaze is usually to increase the strength. When flexure occurs, the maximum stress is found at the surface, so that any flaws in the surface may give rise to fracture. The surface can be improved in this respect by flame polishing, or by creating a compressive surface layer by chilling, but glazing accomplishes two things at least. The glaze tends to heal flaws, but if the thermal expansion of body and glaze are suitable there will be a layer in substantial compression on the surface so as to resist the effect of imposed stresses. On bodies liable to be subject to moisture expansion, compression in the glaze also tends to prevent crazing. For this to be so the thermal expansion of the body must be higher than that of the glaze. When the glaze is in compression there is, of course, a corresponding tension in the body. The magnitude of these stresses will depend on the elastic constants, the body and the glaze thickness, and the thermal expansion differential.

Consider the case of a thin slab of infinite extent. Assume that the body and glaze have the same modulus of elasticity, $E$; that the difference in thermal expansion coefficient is $\Delta\alpha$; that the difference between room temperature and the softening point of the glaze is $\Delta t$; and that the thickness ratio of glaze:body is $r$, then the stresses in the cold material are

for the glaze, in compression $$\sigma_g = -E\Delta t\,\Delta\alpha\,(1 - 3r + 6r^2) \tag{15.3}$$

and

for the body, in tension $$\sigma_b = E\Delta t\,\Delta\alpha r\,(1 - 3r + 6r^2) \tag{15.4}$$

For another simple geometrical shape, a cylindrical rod, let the cross-sectional area of the body be $A_b$ and that of the glaze $A_g$, and $\mu$ = Poisson's ratio. Then the stresses are

for the glaze, in compression $$\sigma_g = \frac{-E\Delta t}{1-\mu}\;\frac{\Delta\alpha A_b}{A_b + A_g} \tag{15.5}$$

and

for the body, in tension $$\sigma_b = \frac{E\Delta t}{1-\mu}\;\frac{\Delta\alpha A_g}{A_b + A_g} \tag{15.6}$$

If we assume the values $E = 7 \times 10^{10}\ \mathrm{Nm^{-2}}$ and $\Delta t = 500\ °\mathrm{C}$, we can calculate the stresses for various values of $\Delta\alpha$ and $r$ from equations (15.3) and (15.4). The results are shown in Fig. 15.4. For the cylindrical rod, taking the value of $\mu$ as 0.25, the stresses are as shown in Fig. 15.5.

It will be seen that the compression in the glaze increases for very thin glazes. Under practical conditions it would be common to find values of $\Delta\alpha = 2 \times 10^{-6}$ and $r = 0.1$ or less, and this would correspond to a compressive stress in the glaze of the order of $5 \times 10^7$ $\mathrm{Nm^{-2}}$. As a consequence, increases in strength of the glazed piece, compared with the biscuit, of around 30 per cent are often achieved. There are, however, some limitations. Excessive glaze compression causes the glaze to break up; a phenomenon known as shelling or peeling. On the other hand, especially with thicker glazes, there is an appreciable amount of tensile stress in the body. If the body contains flaws, fracture may be initiated internally instead of at the glaze surface. Examination of broken rods suggests that when unglazed nearly all fractures start at the surface, whereas

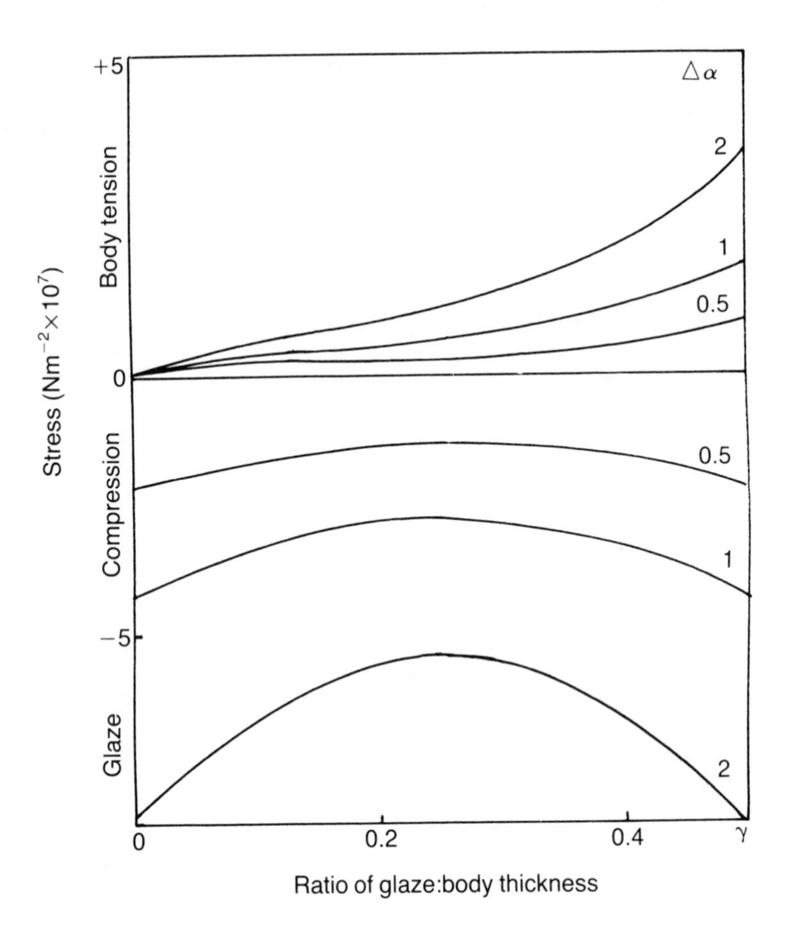


Fig. 15. 4 – Stress in glaze and body due to thermal expansion difference – thin slab.

when glazed perhaps 80 per cent of the fractures start in the interior.

The assumption that the body and glaze have identical elastic properties is not strictly valid; some glazes are much stronger than others. Rado (1971), discussing the properties of hard porcelain, has pointed out that the glaze normally used on this body contains no lead or boron, and is high in alumina and silica, compared with those used on earthenware or china. The properties of the glaze are likely to be significantly different.

Another factor that sometimes complicates the picture is the presence of an intermediate layer between the body and the glaze, either due to excessive reaction between the two or by the deliberate use of an engobe. When this layer has a thermal expansion higher than that of either the body or the glaze, there can be an appreciable loss of strength.

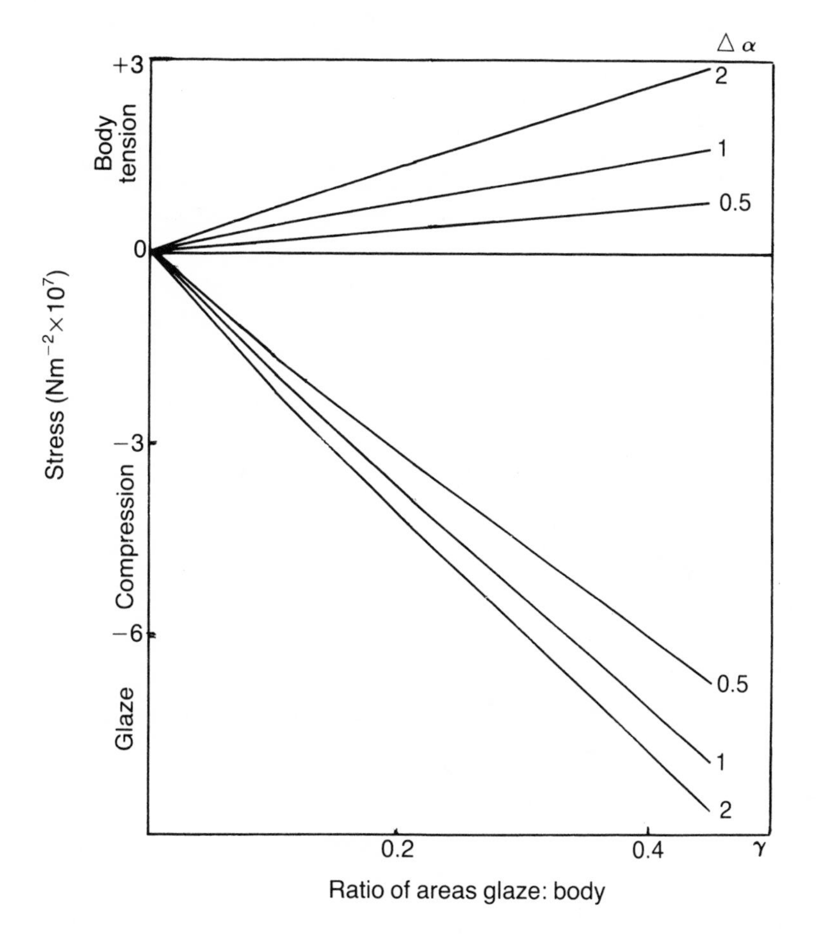


Fig. 15.5 – Stress in glaze and body due to thermal expansion difference – cylindrical rod.

## 15.9 TEMPERATURE

Davies and Brough (1963) investigated the variation of modulus of rupture with temperature. Their results have been summarized in Fig. 15.6, in which the ratio of strength at temperature to the room temperature strength is plotted against the temperature. Above 600 °C the results are ambiguous as the bodies begin to show some viscoelastic behaviour. In the case of a sanitary earthenware body, (A), with quartz as the filler, the strength only begins to increase appreciably at the $\alpha$-$\beta$ quartz inversion region. When the quartz was replaced by cristobalite, (B), a similar increase in strength was observed, but most of it now occurred in the region of the cristobalite inversion. It would seem that as the temperature is raised the effect of the thermal mismatch between matrix and filler is reduced; the structure tightens up, and both Young's modulus and the strength increase. In the case of bone china, (C), there was a decrease in strength to about 500 °C, and then a recovery. The elastic properties of the bone china system are very distinctive, as other parameters, such as internal friction, also reveal. The aluminous porcelain, (D), with very little crystalline inversions involved, shows a fairly steady increase in strength throughout the temperature range.

These data are of no great significance in relation to the service behaviour of whiteware products, but they are of some importance in the study of thermal shock failure, and such problems as dunting during the firing process.

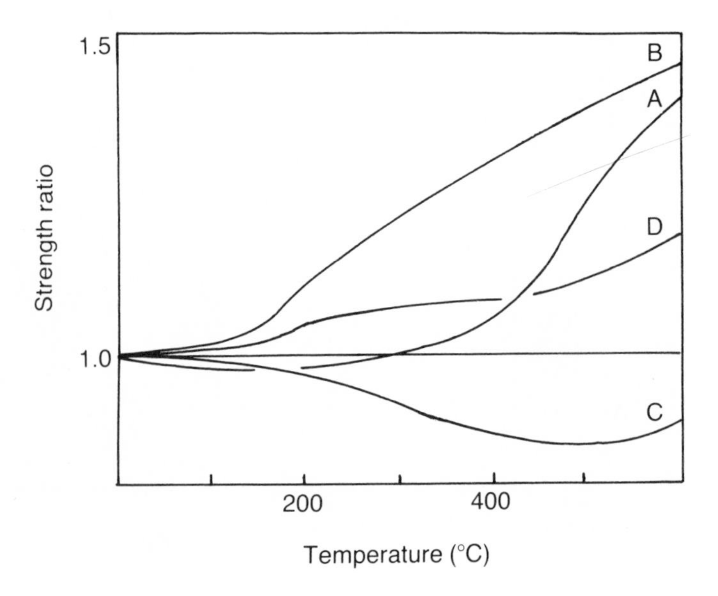


Fig. 15.6 – Variation of modulus of rupture with temperature.

## 15.10 IMPACT

Whiteware products have enough static strength to meet all the requirements of their service life. Bone china, for example, is a very strong material, as may be seen from Fig. 15.7. The fact that it is often regarded as fragile by the general public is a reflection of the thin cross-section often used in tableware. However, all these products are liable to be subjected to impact. In sanitary ware, there is the heavy object dropped into a bowl; there is traffic collision on a tiled wall; and there is the dropping of a cup or plate on a hard floor. How do these bodies react to these mechanical insults? It is well known that the strength of ceramic materials is time dependent, in the sense that they can withstand a high stress for a short time, whereas a lower stress over a longer time can cause fracture. The impact strength is a very important parameter in the field of engineering ceramics, where very high velocities may be encountered. In the present context, however,

Fig. 15.7 – Double-decker bus supported by six bone china cups. (Courtesy Wedgwood.)

we are concerned with a much lower velocity range, of the order of 1-10 $ms^{-1}$. Even this may involve a rate of loading some $10^6$ times as great as that normally used in modulus of rupture tests.

### 15.10.1 Material factor

Dinsdale *et al.* (1962) made a detailed study of this phenomenon using cylindrical ceramic rods, falling in an unrestrained manner against a fixed anvil. In this way, the complicated effects of fixed supports and surface notches were eliminated. The impact occurred at the centre of the rod, and fracture normally occurred at the opposite face, under flexural tension. The energy of the falling rod is distributed in a rather complicated manner between bending, shear and indentation. Measurements were also made, on the same bodies, of modulus of rupture and Young's modulus. The most important finding was that for a wide range of body materials the impact strength was proportional to $S^2/E$, where $S$ is the static modulus of rupture and $E$ is Young's modulus; this ratio being related to the elastic energy used up in bending. Since it is known that these materials tend to fail at a constant strain of about 0.1 per cent, $S/E$ is constant. It is thus found that a plot of impact energy against $S$ is a straight line. The practical significance of this is that, at least for a simple shape such as a rod, impact strength can be predicted from the static modulus of rupture value; in fact, the impact test is merely a rapid modulus of rupture test.

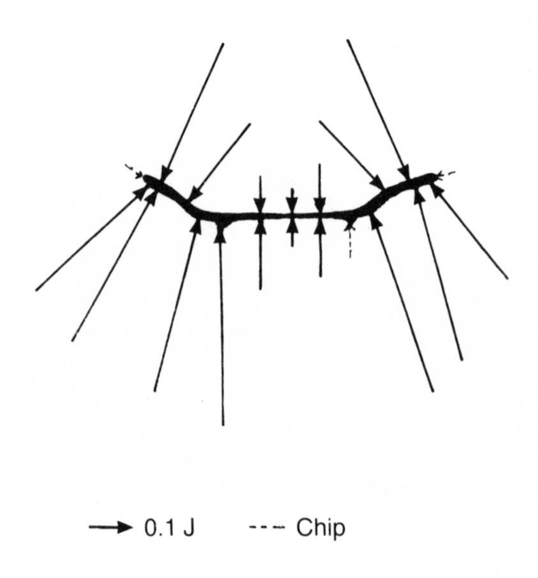


Fig. 15.8 – Energy required to produce fracture in freely suspended plate.

### 15.10.2 Shape factor

In addition to the material factor, the ability of an article to resist impact damage is dependent on shape. Dinsdale *et al.* (1967) investigated the shape factor with particular reference to tableware. Eliminating reactions from rigid supports, they used a freely swinging ball in collision with a freely suspended article. With circular discs they found that the theoretical prediction that impact strength was proportional to the square of the thickness was nearly confirmed, but that the thicker articles had slightly higher strengths than predicted. Curved discs showed that a much higher energy was required to produce fracture when struck on the convex face compared with the concave face, the flat disc falling between the two. Experiments with a dinner plate, struck from many different directions, indicated the need to allow for rotational energy, but also showed how important was the profile of the plate. From these results is is possible to construct a 'survival' envelope such as that shown in Fig. 15.8. In this diagram, the length of the arrows is proportional to the energy required to produce fracture when the plate is struck normally to the surface at that point. Clearly the plate is much more vulnerable from blows in certain directions than in others. Unless something is known about the likely distribution of blows encountered in service, it is thus not possible to be certain about the probable service life of the plate. It may also be mentioned that test procedures involving impact at one point only must be of very limited value.

## 15.11 EDGE CHIPPING

The fracture of the edge of a cup or plate, known as chipping, is of great significance in the service life of tableware products. The edge is most likely to be exposed to blows, and is most vulnerable. Initial damage to the glaze may result eventually in removal of a piece of body and glaze. For thicker pieces the blow required to produce damage will be greater, but the piece removed will be correspondingly larger. The factors that affect the probability of damage can be studied by striking the edge from different directions and measuring the energy required to produce failure.

It is found that the chipping resistance increases rapidly with the thickness of the edge. For plates the extent of the rim is important, larger rims giving some flexibility and hence less chipping; the plate diameter, and thus the rim curvature, does not have much effect. Much the most important factor is the radius of the edge in the opposite plane. Sharp edges chip easily; edges with a large radius of curvature are much stronger. Differences in chipping resistance of up to 30 to 1 can be found with edges of different shapes, overshadowing any changes that can be brought about by improved body strength.

The practical difficulty is that it is not easy to produce an edge profile of the required curvature with present methods of towing and edge fettling. Means of attaining consistency in this regard would yield dividends in terms of service durability. The importance of this subject is underlined by the fact that studies of damage in use show that chipping occurs much more frequently than total fracture.

Fig. 15.9 shows the measured chipping resistance for a number of edge shapes; the three on the left accurately formed on discs, the two on the right being plates from factory production.

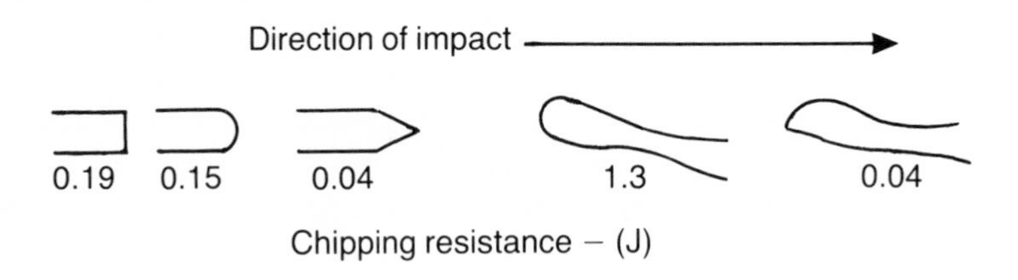


Fig. 15.9 – Effect of edge profile on chipping resistance.

## 15.12 CONCLUSION

The strength of whiteware products is an important element in their fitness for function. In general, it is adequate, but it can be improved by attention to body composition, firing, and correct glazing. Nevertheless, the practical improvements are never more than about 2:1, and are reflected in roughly that ratio in service life. In the particular case of edge chipping, shape is of paramount importance. Design is also critical in complicated shapes like porcelain insulators, where stress distribution patterns may overturn material factors.

## REFERENCES

Batchelor, R. W. and Dinsdale, A. (1960) *Transactions VII International Ceramic Congress,* 31.

Beech, D. and Norris, A. W. (1961) *Tr. Brit. Cer. Soc.,* **60**, 556.

Binns, D. B. (1962) *Sci. of Ceramics,* **I**, 315.

Davies, L. J. and Brough, R. (1963) *Tr. Brit. Cer. Soc.,* **62**, 463.

Dinsdale, A., Moulson, A. J. and Wilkinson, W. T. (1962) *Tr. Brit. Cer. Soc.,* **61**, 259.

Dinsdale, A. and Wilkinson, W. T. (1966) *Proc. Br. Cer. Soc.,* **6**, 119.

Dinsdale, A., Camm, J. and Wilkinson, W. T. (1967) *Tr. Brit. Cer. Soc.,* **66**, 367.

Rado, P. (1971) *Tr. & J. Brit. Cer. Soc.,* **70**, 131.

# 16

# Properties of finished ware — thermal

The thermal properties of ceramic materials in general are of great interest to designers of engineering components for use at high temperatures, or in conditions of severe thermal shock. In fact some materials are found to be unacceptable solely on the grounds that they fail to meet the stringent criteria, though many of their other properties may be favourable.

In our present context, however, we are concerned with a more limited aspect of the topic. The thermal properties of fired whiteware products are relevant in two main areas. The first is in the matter of firing. Thermal stresses set up during the firing process, especially during cooling, can give rise to serious losses. Further, the rate at which ware can be heated and cooled determines the limits to the possibility of fast firing, and a knowledge of the fundamental thermal properties is essential here. The second area relates to the performance of the product in use. All whiteware products may be subject to a degree of thermal shock in their normal use; at this level thermal shock resistance is a complex property, difficult to either define or measure. Apart from thermal shock, the ability to transmit or retain heat is important. Tea in teapots or drinking vessels, and hot water in sanitary applicances, are examples that come readily to mind.

In some areas data are sparse for whiteware bodies, but the basic principles involved can still be identified and examined.

## 16.1 SPECIFIC HEAT CAPACITY

When a ceramic material is heated, the amount of energy required to produce a given rise in temperature is characteristic of the material. The energy is required for a number of purposes: to provide for the increased amplitude of the vibrations of atoms in the lattice, rotational energy, and, to a smaller extent, changes in defect structures or electronic configurations.

For any given substance, there is a parameter that gives a measure of this heat requirement. It used to be known as the specific heat, and was defined as the quantity of heat required to raise the temperature of one gram of the substance through one degree centrigrade. The unit was cal.$g^{-1}$.° $C^{-1}$, and the value for water was 1. In modern terms the constant is known as the specific heat capacity, and is denoted by $c$. The unit is J $kg^{-1}$ $K^{-1}$, and the value for water is 4185. Strictly speaking, there are two specific heat capacities, relating to constant pressure and constant volume, but the difference between the two is of no significance here.

Specific heat capacity varies not only with the nature of the material, but also with temperature, and there are some important relationships that we need to note. Dulong and Petit's law states that the atomic heat capacity, that is the

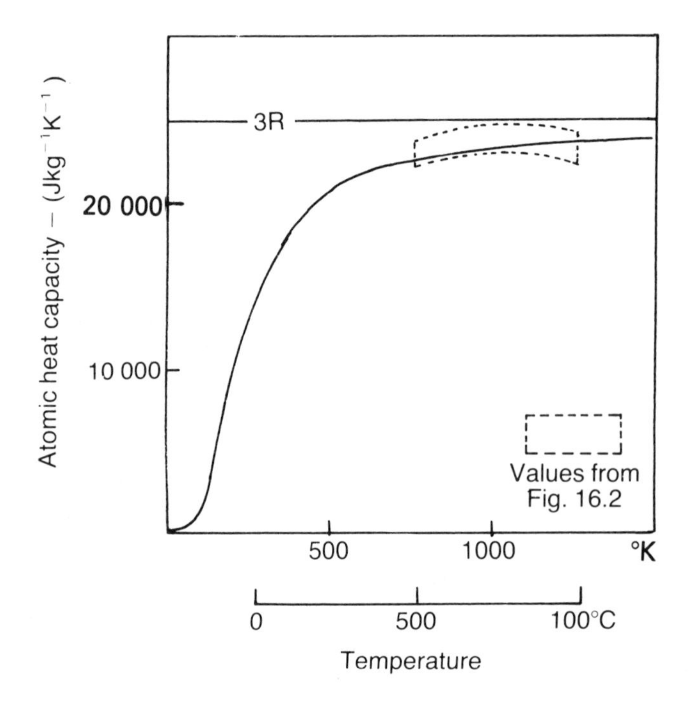


Fig. 16.1 – Variation of atomic heat capacity with temperature.

specific heat capacity multiplied by the atomic weight, is the same for all substances, and should tend towards a constant value at high temperatures equal to three times the gas constant. The limiting value is thus 5.96 calories per g atom, or 24 940 J per kg atom. It turns out in practice that this law is a good representation of the facts at high temperatures, both for elements and compounds. In the case of compounds it is necessary to use a mean atomic weight. There are some differences, due to the fact that real substances are very different from simple crystals; nevertheless, the general concept holds good.

The variation of specific heat capacity with temperature is complex, but here again a common pattern is found. For a wide range of substances the curve is of the form shown in Fig. 16.1. At very low temperatures the heat capacity is proportional to the cube of the absolute temperature, $T$. Different substances have curves of the same shape, but displaced from each other. They can be reconciled by the use of a temperature, $\theta$, characteristic of the substance, known as the Debye temperature. When the heat capacities are plotted as a function of $T/\theta$, the curves for different substances are identical. Over the middle range it is often found that the specific heat capacity varies with temperature according to the equation

$$c = a + \mathrm{b}T - \mathrm{d}T^2$$

With regard to absolute values, some data published by Clements (1962) for a series of ceramic materials, have been recalculated and plotted in Fig. 16.2.

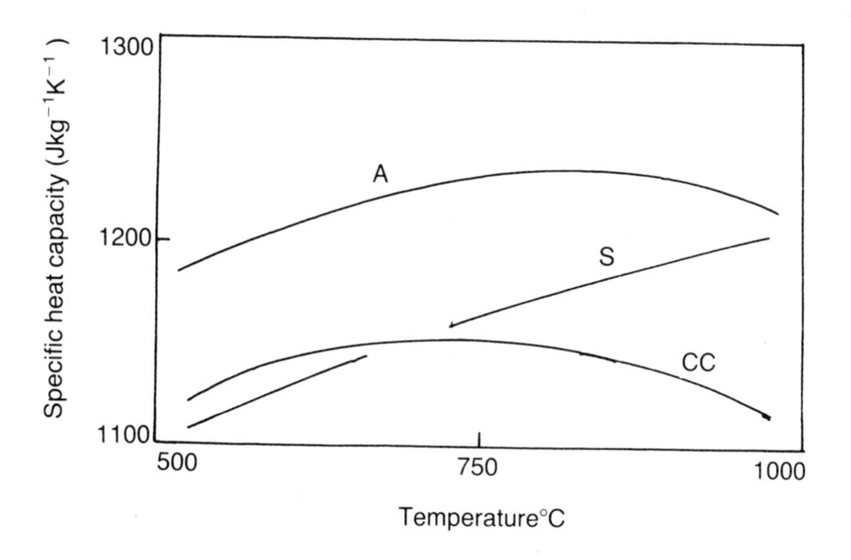


Fig. 16.2 – Specific heat capacities for ceramic materials. A, 95% alumina; S. silica brick, CC, fired china clay.

These values can be transformed into atomic heat capacities by multiplying by a mean atomic weight. It is interesting to note here that the mean atomic weight of $SiO_2$, 20, is very close to that for $Al_2O_3$, 20.4, so these two oxides have virtually the same heat capacity. The data in Fig. 16.2, expressed in terms of atomic heat capacity, are shown in Fig. 16.1, where the area is seen to fit reasonably well on the theoretical curve. The concept of additive atomic heats should not be pressed too far, but Kopp's rule uses a number of factors for this purpose.

Measured values for whiteware products are sparse and somewhat variable. There is probably not much difference between different bodies, and all would probably be around 800 $J\,kg^{-1}\,K^{-1}$ at room temperature, and 1200 at 1000° C. Two other points need to be noted. All the values quoted here relate to true specific heat capacities, that is the heat required over one degree at a specified temperature. In many industrial applications, when it is required to compute total heat inputs over a range of temperature, so-called interval heat capacities are used. This figure is derived from the total heat divided by the temperature interval. Data for true and interval specific heat capacities can be significantly different numerically, and care should be taken to distinguish between them. In the second place, it should be noted that we are here concerned only with the data for fired products. The heat requirement in heating up clay-based bodies includes that involved in the reactions taking place, and the effective specific heat capacity is accordingly different. This aspect has been dealt with in Chapter 11.

## 16.2 THERMAL EXPANSION

Most substances, though not all, expand when heated. The reasons why some substances either contract, or show no volume change on heating, are beyond the scope of this book. For whiteware products expansion is the norm. It is important from the point of view of glaze fit, dealt with in Chapter 13, and also in connection with the development of thermal stresses. When a crystalline substance is heated, the amplitude of the vibrations of the constituent atoms is increased. Because the effect of this on the repulsive forces between atoms is different from the effect on the attractive forces, there is a shift of the equilibrium position reflected in a volume change. If it were not for this inharmonicity, no expansion would occur. As the same process is involved in the change of specific heat capacity with temperature, it is not surprising to find that the ratio of specific heat capacity to thermal expansion coefficient is constant at the same temperature, as stated in Gruneisen's law. In practical applications, then, note must be taken of the fact that the expansion coefficient increases with temperature.

The coefficient of linear thermal expansion, $\alpha$, is defined as the change in length per unit length per unit change in temperature. The unit is thus $K^{-1}$. It should be noted here that anisotropic crystals expand to different degrees in different directions. Also, for isotropic substances, the coefficient of volume expansion may be taken as three times the linear coefficient, for most practical purposes.

Expansion curves were given for some typical whiteware bodies in Fig. 13.3. Here the total expansion to a stated temperature is shown; in connection with glaze fit, a useful figure to quote is the total expansion to 500° C, which varies between 0.2 and 0.4 per cent. The true coefficient of expansion at any temperature is found from the slope of the curve. These values are shown in Fig. 16.3. It will be seen that a body containing a lot of quartz shows the effect of the volume change due to the $\alpha$-$\beta$ quartz inversion around 500° C. Similarly, body A shows the effect of the cristobalite inversion in the 200-300° C range. Bone china shows a more gradual increase of $\alpha$ with temperature, although even there the influence of small traces of quartz can be seen. The degree of solution of quartz in the glassy matrix can cause appreciable changes; in high temperature porcelains, for example, it can reduce the expansion by half. It is greatly affected by the grain size of the quartz, and other factors which have been studied by Schuller and Lindl (1964).

Some typical values for thermal expansion are given in Table 16.1. In practice, values may range through ±15 per cent of these values. The corresponding mean linear coefficient of expansion over this temperature range thus varies from 3 to $8 \times 10^{-6}$ $K^{-1}$.

The localized effect of volume changes due to crystalline inversions has been dealt with in Chapter 15, with special reference to strength. However, thermal stresses can be set up in isotropic materials, free from discontinuities, due to

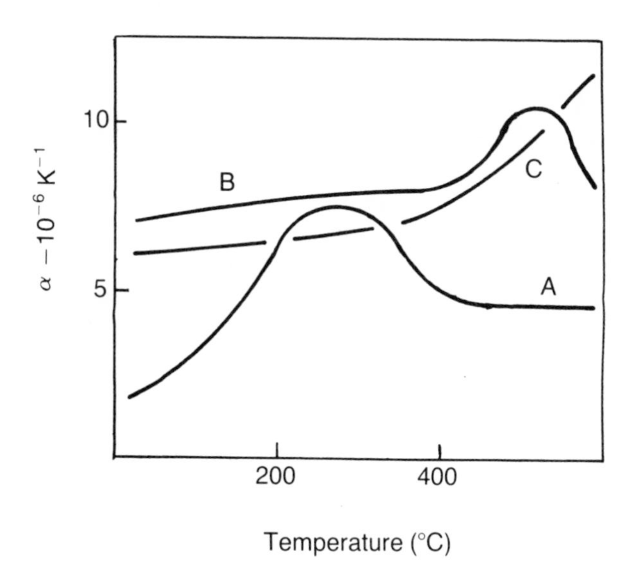


Fig. 16.3 – Linear thermal expansion coefficients for various bodies. A, Body high in cristobalite; B, bone china; C, body high in quartz.

**Table 16.1** – Linear percentage thermal expansion over the range 0–500° C

| | |
|---|---|
| Bone china | 0.42 |
| Earthenware | 0.38 |
| Vitrified ware | 0.32 |
| Porcelain | 0.25 |
| Cordierite porcelain | 0.15 |

differences in temperature between adjacent elements in the body.

Consider a rod of material restrained at the two ends, and let the temperature change be $\Delta t$. Since the rod is not allowed to expand, a stress will arise, equal to the product of Young's modulus $E$ and the strain $\epsilon$. Then, if $\alpha$ is the coefficeint of thermal expansion, we have $\epsilon = \alpha\Delta t$ and $\sigma = E\epsilon = E\alpha\Delta t$. If we assume that the rod is being cooled, and hence in tension, and that the critical breaking strain in tension is $10^{-3}$, then for a body with $\alpha = 5 \times 10^{-6}\ \text{K}^{-1}$, failure would occur with a temperature change of $\Delta t = \epsilon/\alpha = 10^{-3}/(5 \times 10^{-6}) = 200°\ \text{K}$.

Whether such a temperature gradient is likely to occur or not in a given material depends on its response to a changing thermal environment, and, in particular, how quickly it dissipates heat. This leads us to consider two other important thermal properties involved in heat transfer, namely thermal conductivity and thermal diffusivity.

## 16.3 THERMAL CONDUCTIVITY

We consider first the conduction of heat through a body under steady state conditions, that is with the temperature distribution constant with respect to time. The rate of heat flow at any point is proportional to the temperature gradient.

Thus, if we consider a slab of thickness $x$, with a uniform temperature gradient $\mathrm{d}t/\mathrm{d}x$, the rate of heat transfer through an area $A$, is given by

$$\frac{\mathrm{d}Q}{\mathrm{d}\theta} = A\ \frac{\mathrm{d}t}{\mathrm{d}x}$$

$\leftarrow l \rightarrow$

$\vec{Q}$ $t_1$ $t_2$

The quantity of heat $Q$ flowing per second is thus

$$Q = \lambda . A .\ \frac{t_1 - t_2}{l}$$

The constant $\lambda$ is known as the thermal conductivity; it is the amount of heat passing unit area, in unit time, under a unit temperature gradient. It is a characteristic of the particular material, and the unit in which it is measured is $J\ s^{-1}\ K^{-1}\ m^{-1}$.

Since 1 Joule per second equals 1 Watt, the unit can also be expressed as $W\ K^{-1}\ m^{-1}$. The value for whiteware ceramic materials is approximately between 1 and 1.5. This may be compared with approximate values of 0.2 for wood, 0.3 for plastics, 0.6 for water, 3 for iron, and 32 for copper. The practical implications of these values may be seen by considering, for example, the retention of heat in hot water in kitchen and bathroom appliances. Since the rate of heat loss through the walls of a vessel is proportional to $\lambda/x$, it will be seen that for a metal teapot, with $x = 1$ mm and $\lambda = 10$, a ceramic teapot, with $x = 4$ mm and $\lambda = 1.25$, and a ceramic wash-basin, with $x = 12$ mm and $\lambda = 1.25$ the rates of heat loss are in the ratio 100:3:1.

Conduction, however, is not quite the simple process that the above analysis suggests. It is actually very complex, and, unlike specific heat, is greatly influenced by the texture of the material. For solid substances in general, there are two main mechanisms available for the transmission of energy. In one, the energy is transferred by the movement of free electrons, and this is the main process in the case of metals. In the case of insulators, such as most ceramics, where there are practially no free electrons, transfer is by phonons, or quanta of energy, through the vibrations of atoms in the crystal lattice. Intermediate between the two are semi-conductors, in which both processes operate; silicon carbide is a good example. With perfect single crystals, such as alumina, the thermal conductivity is high, and its variation with temperature is predictable by theory. At very low temperatures, $\lambda$ decreases according to $1/t$; at very high temperatures, $\lambda$ is constant. At high temperatures, heat transfer by photons, becomes important; less so at lower temperatures since radiation varies with $t^4$.

Any departure from crystallinity results in a lowering of $\lambda$. Highest values are found with simple crystals, of low molecular weight. The inclusion of atoms of higher atomic weight reduces $\lambda$. Indeed, anything that impedes the flow of energy has this effect, and this includes defects, pores, and grain boundaries. The effect of grain boundaries is seen in the fact that the conductivity of the polycrystalline form of an oxide such as alumina falls increasingly below that of the single crystal of the same material as the temperature increases. Within the range usually found in whiteware bodies, the effect of grain size is not important. The irregular structure of glass results in low conductivity. Thus, the contribution of the crystalline components in whiteware bodies is diluted by the glassy matrix.

Porosity is important, which is why refractory insulating materials are generally porous. The surfaces of the pores act as barriers to the transmission of energy. Since large pores have fewer interfaces per unit length of solid, conductivity is less affected by large pores than by small ones. For whiteware bodies, we may say that $\lambda$ decreases with increased porosity, but at any given level of porosity, $\lambda$ increases with pore size.

Temperature dependence varies with the absolute value of $\lambda$. For most materials with a high value of λ, the coefficient is negative. For low values of λ, the coefficient tends to be positive, and this is the case with whiteware bodies, where λ may be approximately doubled from 0 to 1000° C. For most aspects of the performance of whiteware bodies in service the value in the range 0–100° C is the most significant. Some approximate values for whiteware bodies are:

| | |
|---|---|
| Earthenware and fireclay | 1.25 $W\,m^{-1}\,K^{-1}$ |
| Porcelain and vitreous ware | 1.7 |
| Alumina porcelain | 4 |
| Alumina | 15 |

## 16.4 THERMAL DIFFUSIVITY

We consider now the flow of heat in the unsteady state in which temperature varies with both position and time. This is of much greater practical significance than the steady state condition, being concerned with the response of a solid to changes in its thermal environment which may be sudden, gradual or cyclic. Transient heat flow is similar to many other diffusion processes and can be represented by a Fourier equation, relating to flow in one dimension only,

$$\frac{\partial t}{\partial \theta} = \frac{\lambda}{c\rho} \frac{\partial^2 t}{\partial x^2} \tag{16.1}$$

where $t$ is the temperature after time $\theta$ at a point distant $x$ from a reference plane, and λ is the thermal conductivity, $c$ the specific heat at constant pressure and $\rho$ the density. The quantity $\lambda/c\rho$ is called the thermal diffusivity, which we represent by the letter $a$. It is not in itself a physical quantity; its significance is that it is an indication of how much heat is retained by an element of the solid, and how much is transmitted to an adjacent element. The unit of thermal diffusivity is $m^2\,s^{-1}$. It can be measured directly by the flash method, in which a heat pulse of short duration is applied to one face of a laminar specimen, the rate of change of temperature being measured on the opposite face.

An estimate of its approximate value for whiteware bodies can be made. Taking $\lambda = 1.5\ J\,s^{-1}\,m^{-1}\,K^{-1}$, $c = 800\ J\,kg^{-1}\,K^{-1}$ and $\rho = 2500\ kg\,m^{-3}$, we have

$$a = \frac{\lambda}{c\rho} = 0.75 \times 10^{-6}\ m^2\,s^{-1}.$$

which is in good agreement with such published data as are available.

The influence of such factors as porosity and temperature on $a$, can be inferred from what has already been described in connection with λ, $c$ and $\rho$ individually. For example, $c$ and λ both increase with temperature, while $\rho$ decreases. It might be expected, then, that $a$ would increase with temperature.

Values published by Holmes (1969) show that $a$ for bone china increases from $0.5 \times 10^{-6}$ m$^2$ s$^{-1}$ at room temperature to $0.85 \times 10^{-6}$ m$^2$ s$^{-1}$ at 1000° C.

If $a$ varies with temperature, the solution of equation (16.1) becomes difficult, but much can be learnt from examining the solution when $a$ is constant. Consider a slab of material of infinite extent, initially at a temperature $t_i$, and suppose the temperature at the surface to be instantaneously changed to $t_s$. The solution of equation (16.1) would relate the temperature $t$ at any point $x$ after time $\theta$ by the expression

$$\frac{t_s - t}{t_s - t_i} = f\left(\frac{x}{2\sqrt{a\theta}}\right) \tag{16.2}$$

The function $f(\ )$ is the well known Gauss error integral and can be obtained

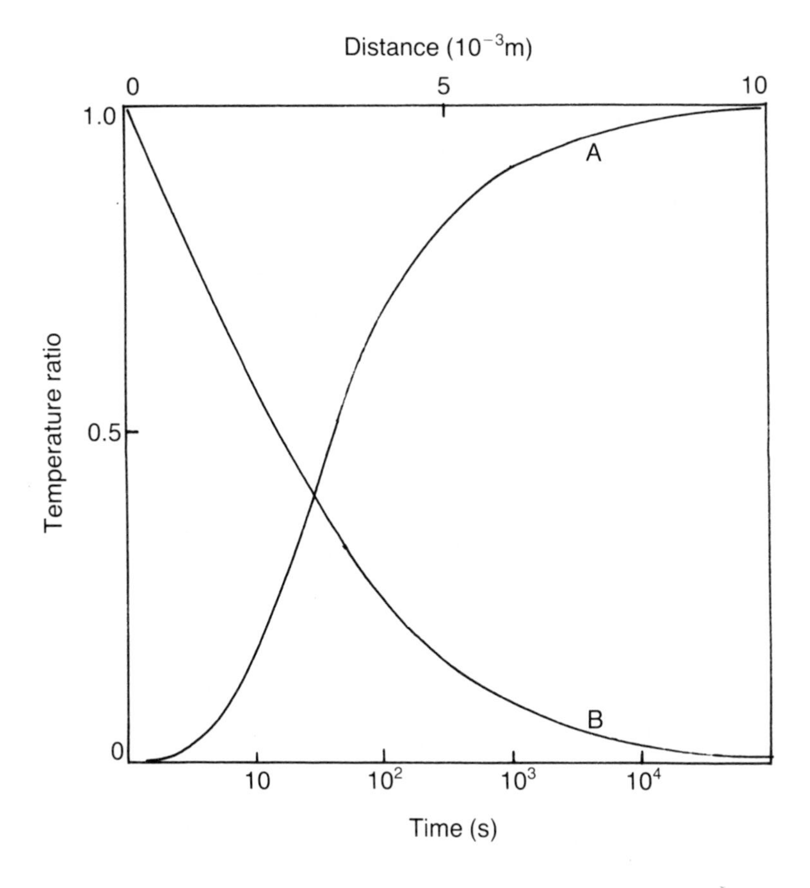


Fig. 16.4 – Temperature at distance $x$ from surface after time $\theta$. Curve A, constant position; curve B, constant time.

from standard tables. The function is plotted in Fig. 16.4. The ordinate shows the temperature as a fraction of the initial temperature difference between surface and interior. In curve A, we see how the temperature increases with time at a fixed position. Curve B shows how the temperature at a given time decreases away from the surface. The shape of the curves shown is independent of the dimensions of the material, since $x/2\sqrt{a\theta}$ is a dimensionless quantity. It is important to note that the temperature varies as the inverse square of the distance, a fact which is of importance when considering some practical issues. The gradient of curve B, $dt/dx$, is a measure of the thermal stress likely to be set up under these circumstances, and it will be noted that it is a maximum at the surface. This means that probable thermal shock failure can be estimated in terms of the maximum surface stress.

If we consider a slab of finite thickness suddenly cooled at the surface, we can calculate the stress distribution throughout the thickness, with the result shown in Fig. 16.5. The centre of the slab is in compression and the surface in tension. If the surface were to be suddenly heated, the reverse would be the case. Since ceramics fail much more readily in tension than in compression, it

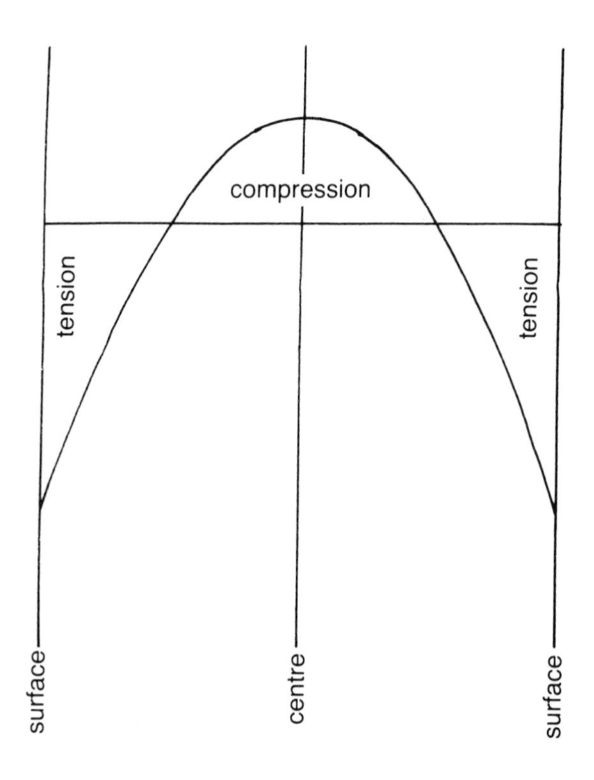


Fig. 16.5 – Stress distribution in a plate cooled from the surface.

will be appreciated that cooling is more likely to produce thermal shock damage than is heating. It will be seen that the stress distribution is parabolic, and that the maximum stress occurs at the surface.

The magnitude of the maximum stress depends on the geometry of the solid, but can be calculated for simple shapes that represent to some degree cases likely to be encountered in practice. The stress depends on the difference between the surface temperature, $t_s$, and the average temperature throughout the interior, $t_a$, as well as on the thermal expansion $\alpha$, Young's modulus $E$, and Poisson's ratio $\mu$.

| | Maximum surface stress $\sigma$ |
|---|---|
| Thin plate | $(t_a - t_s)\,E\alpha$ |
| Thick plate, cylinder or sphere | $(t_a - t_s)\,E\alpha/(1-\mu)$ |
| Thin disc | $(t_a - t_s)\,E\alpha.(1-\mu)/(1-2\mu)$ |

We can make an estimate of the stress, by inserting typical values. If a bone china plate, which would be represented by a thin disc, were heated to 100° C and suddenly plunged into water at 0° C, the stress would be given by

$$\sigma = \frac{100\,E\alpha/(1-\mu)}{(1-2\mu)}$$

Taking $E = 10^{11}$ $\mathrm{Nm}^{-2}$, $\alpha = 8 \times 10^{-6}$ $\mathrm{K}^{-1}$ and $\mu = 0.25$, we have

$$\begin{aligned}\sigma &= 100 \times 10^{11} \times 8 \times 10^{-6} \times 1.5\\ &= 1.2 \times 10^{8}\ \mathrm{N\,m^{-2}}\end{aligned}$$

This is of the same order as the measured strength of bone china, but, as we shall see later, the risk of failure is reduced by the fact that the surface does not immediately take up the temperature of its surroundings.

## 16.5 THERMAL SHOCK RESISTANCE

Risk of thermal shock damage may be encountered with whiteware bodies in two main areas. In the firing process, where stresses may be set up at rapid rates of cooling, the phenomenon of dunting is well known to potters. So far as products in use are concerned, the most likely thermal shock occasion is contact with boiling water. Cooking ware is an exception.

Much research has been carried out in recent years on the thermal shock resistance of ceramics. In engineering components and furnace structures at high temperatures it is of great importance. In these fields, much attention has been given to crack propagation, a distinction being made between crack initiation and final failure. With whiteware products, this distinction is generally irrelevant. For a limited number of materials, the presence of large pores may mean that a thermal crack terminates at a pore. This is certainly the reason why the red-clay teapot could be placed on a hob adjacent to a coal fire without catastrophic damage. But for most whitewares, and especially vitreous wares, the initial crack means failure.

Many attempts have been made to establish a thermal shock resistance factor, both by experiment and theory. Experiments on actual ware can be carried out on small articles, such as cups and plates. For large pieces, such as sanitary ware, resort has to be made either to test pieces, or small-scale models of the shape. Experimental results do not scale up with any degree of certainty.

Attempts are often made to derive a thermal shock resistance factor from the involved parameters. Thus the temperature difference, $\Delta t$, required to produce failure, is taken as a measure of thermal shock resistance, and

$$\Delta t \ \alpha \ \frac{\text{strength}}{\text{Young's modulus} \times \text{thermal expansion}}$$

Quantitative correlation with experimental data is never better than approximate, for a number of reasons. In the first place, there is the assumption that fracture will occur at a constant strain, but in fact the critical strain varies both with the material and the tempeature. Secondly, it is necessary to introduce a shape factor into the above equation. But, thirdly, and most importantly, it is difficult to define the conditions of heat transfer at the surface. A sudden change in temperature of the environment is not immediately reflected in the temperature of the surface layer of the solid. The rate of response depends on a surface heat transfer factor, $h$, which may have very different values for different conditions. For instance its value, in $\text{J m}^2 \text{ s}^{-1} \text{ K}^{-1}$, may vary from 10 to $10^3$ in air, according to velocity, and from $10^3$ to $10^4$ in a water quench. A parameter known as Biot's modulus, $\beta$, is often introduced into these calculations. It is equal to $h/\lambda \times$ a dimension, and is thus dimensionless. It enables a quantitative assessment to be made of the importance of dimension, for example thickness of ware, for given values of $h$ and $\lambda$.

The theoretical treatment, then, cannot be more than a guide. It is certainly true, in general terms, that good thermal shock resistance is a function of high strength and low thermal expansion. Results of experiments on ware, however, do give a practical guide to levels of thermal shock resistance to be expected. Wilkinson and Sorrell (1963) reported the results of experiments on bone china cups, following work on cylindrical rods which suggested that no damage was caused by quenching in iced water from temperatures up to 150° C. Using glazed cups, a series of treatments was examined. Cups heated in an oven were plunged into a bath of cold water; cold water was poured into the cup up to various levels; the cups were placed in cold water up to various levels, with no water entering the cup; and 100 cycles between cold and boiling water. No failures were observed at temperatures up to 160° C. It was also found that thermal shock failure could be induced by impact damage to the rim, at a level such that the crack could not be seen by the naked eye, but could be revealed by dye penetration. It may be concluded from these observations that bone china will not suffer thermal shock damage under all normal conditions of use.

Cooking is another field where thermal shock may be an important feature. For direct use in very hot ovens, or on gas or electric rings, special low thermal expansion bodies are needed, such as may be obtained in glass-ceramics. However, in recent times the use of so-called oven-to-tableware has become increasingly popular. This ware may be made of normal tableware body, and glazed, in shapes such as casseroles which can be used for heating food in the oven and serving direct to table. In this case oven temperatures up to 175° C may be encountered, and placing the article directly on a cold surface may represent an appreciable thermal shock. Tests on this type of ware are notoriously unreliable, but do not seem to correlate with ideas about the relevant parameters, such as thermal expansion and strength. What emerges from these considerations is the fact that in the thermal shock resistance formula another factor need to be introduced – the shape factor. As we noted when discussing surface heat transfer, thickness of the body can be a crucial factor, but other features of the design can be equally significant. For casseroles, for example, a protruding rim or foot lowers the resistance, as does a sharp angle at the junction of bottom and sides. In general terms, sudden changes in either thickness or curvature are undesirable. This design factor cannot be quantified, but it is certainly as important as any of the other properties.

## 16.6 FIRING

When ware is heated or cooled too rapidly in the kiln, temperature differences may be set up between different parts of sufficient magnitude to produce failure. For example, if the rim of a plate cools too rapidly compared with the centre, a tangential tensile stress will arise at the edge and a radial crack may occur. Stephenson (1976) has studied this problem in connection with the commercial production of vitrified plates in a tunnel kiln. Calculating critical strain values, and using the stress distribution in a disc as representative of that in a plate, he has related cracking incidence to these factors and to kiln speed. Holmes (1969) has discussed the possibilities for very fast firing, a subject that is of considerable interest at the present time. Using measured values of diffusivity for several whiteware bodies, and solving the diffusivity equation numerically for 99.9 per cent of the equilibrium temperature, he has calculated the temperature gradients as a function of time. He has shown that single pieces of ware, such as a tile or a plate, can tolerate very high rates of surface heating, providing that the heat is applied uniformly, and that one dimension is small. The difficulties increase rapidly when the thickness is increased.

## 16.7 CRITICAL STRAIN

In much that is written about the elastic properties of ceramics, the concept of critical strain has been widely accepted. It is also commonly assumed that it does not vary widely with temperature, but in thermal shock considerations it may be that the variation with temperature is of some significance. Values for a number of interesting whiteware bodies have been calculated from data on Young's modulus and strength published by Davies and Brough (1963). The variation of critical strain with temperature for these bodies is shown in Fig. 16.6. The sudden changes in the case of quartz and cristobalite are clearly shown, but even with the other two bodies the variation is noticeable. Formulae for thermal shock resistance that assume invariability with respect to temperature for critical strain may thus be in some danger of error.

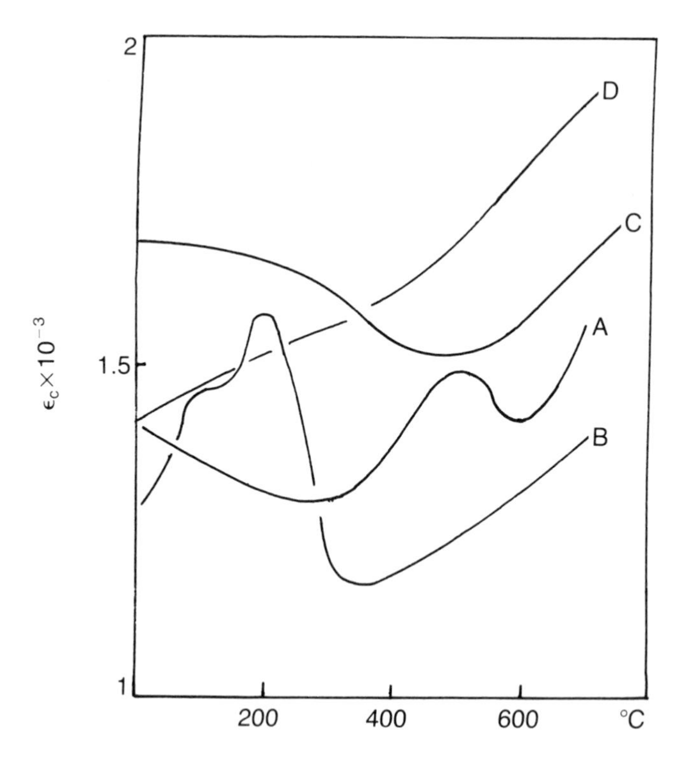


Fig. 16.6 – Variation of critical strain with temperature. A, quartz body; B, cristobalite body; C, bone china; D, aluminous porcelain.

## REFERENCES

Clements, J. F. (1962) *Tr. Brit. Cer. Soc.,* **61**, 452.
Davies, L. J. and Brough, R. (1963) *Tr. Brit. Cer. Soc.,* **62**, 451, 463.
Holmes, W. H. (1969) *J. Brit. Cer. Soc.,* **6**, 19.
Schuller, K. and Lindl, P. (1964) *Intern. Cer. Congress,* IX, 139.
Stephenson, R. J. (1976) *Tr. & J. Brit. Cer. Soc.,* **75**(6), 127.
Wilkinson, W. T. and Sorrell, J. (1963) *Tr. Brit. Cer. Soc.,* **62**, 39.

# 17

# Properties of finished ware — optical

Having considered thermal properties we move down the electromagnetic spectrum, into the region of wavelengths 400–700 nm, and look at optical properties. Important as the technical properties, such as strength, undoubtedly are, most whiteware ceramics are sold on eye-appeal. The customer is interested in aesthetic appearance, and this is true of tiles and sanitary ware as well as for tableware and ornamental products. Aesthetic appeal is, of course, notoriously difficult to define. It is not a physical quantity, and we are here in the realm of psychophysics. Good design has the double purpose of achieving fitness for function, and attractive and pleasing visual impression.

We have dealt with colour in Chapters 13 and 14, and we may mention here the fact that in addition to glazes and decorations, coloured bodies make their appeal. Fashions change as in many other kinds of consumer goods. Sometimes very white bodies are in vogue, and these can now be provided thanks to the many advances in the technology of clay refinement. At other times, creamware is fashionable, and all-over coloured bodies in shades of blue, green, and pink, are invariably popular. All these effects derive from the use of colouring oxides in body, glaze, or decoration recipes.

But there are other features as important as colour, and we deal in this chapter with three of the most important, namely, opacification, gloss, and translucency. Before dealing with these in detail we look at some of the basic relationships that need to be taken into consideration in the propagation of light.

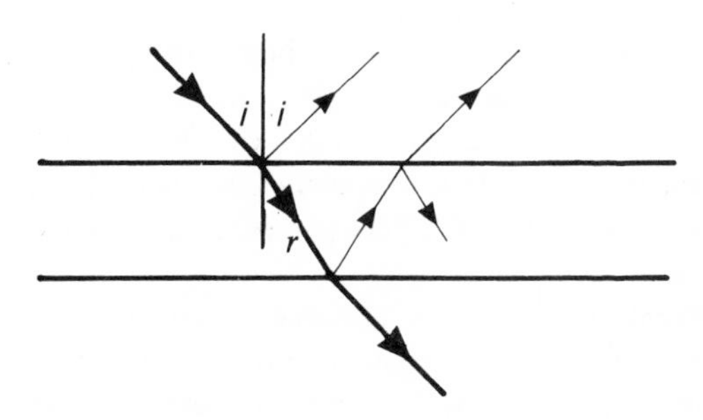


Fig. 17.1 – Reflection and refraction at plane surfaces.

## 17.1 BASIC RELATIONSHIPS

Light is a form of electromagnetic radiation, travelling in straight lines in an isotropic medium. When it enters a dielectric, the velocity is reduced; the ratio of the velocity *in vacuo,* or for practical purposes in air, to the velocity in the dielectric, is known as the refractive index, which we denote by $n$. It is proportional to the square root of the dielectric constant of the material.

As a consequence of this change of velocity, there is a change of direction at a boundary between two media, as shown in Fig. 17.1. Some of the light is reflected and some transmitted. For reflected light the angle of incidence $i$ is equal to the angle of reflection $r$. For refracted light, the directions are related by Snell's law, namely $\sin i/\sin r = n$. Light passing from a more dense to a less dense medium may be totally reflected internally, if the angle of incidence is greater than a critical angle $= \sin^{-1}(1/n)$. Thus, under certain circumstances, multiple reflections and refractions may take place.

It is important to note that at every interface there is a loss of intensity. This can be calculated from the Fresnel equation

$$\frac{2I_R}{I_0} = \left\{\frac{(n^2 - \sin^2 i)^{\frac{1}{2}} - \cos i}{(n^2 - \sin^2 i)^{\frac{1}{2}} + \cos i}\right\}^2 + \left\{\frac{\cos i - \dfrac{1}{n}\left(1 - \dfrac{\sin^2 i}{n^2}\right)^{\frac{1}{2}}}{\cos i + \dfrac{1}{n}\left(1 - \dfrac{\sin^2 i}{n^2}\right)^{\frac{1}{2}}}\right\}^2 \tag{17.1}$$

where $I_0$ is the incident intensity, and $I_R$ the intensity of the reflected beam. For normal incidence, this reduces to

$$\frac{I_R}{I_0} = \left\{\frac{(n-1)}{(n+1)}\right\}^2 \tag{17.2}$$

Taking a value of 1.5 for $n$, Equation (17.1), shows that the intensity of the reflected beam is about 4 per cent of that of the incident beam for small angles of incidence, rising to a value of 100 per cent at glancing incidence. The important thing to note is that every time there is reflection at an interface there is a loss of intensity of a few per cent, an important fact in connection with scattering.

Internal reflection may take place a number of times, as shown in Fig. 17.1. If either the first, or subsequent surface, is rough, light will be reflected in all directions, a feature we will consider in more detail when analysing gloss.

Obstructions in the path of the light such as pores or small crystals result in

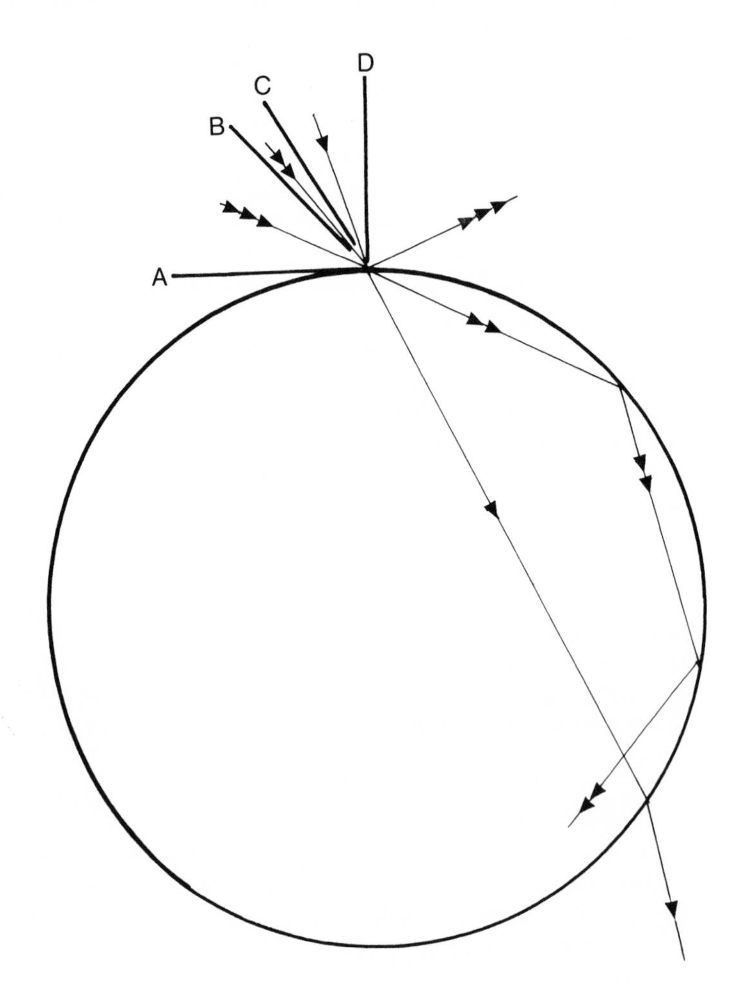


Fig. 17.2 – Scattering of light by a spherical bubble.

a diminution in intensity and in the phenomenon known as scattering. Simple theory, dealing with obstruction only, indicates that the scattering effect depends on the effective cross-sectional area of the particle or bubble, and on the volume concentration. However, the process is much more complex than that, since multiple reflections usually take place. What may happen to a ray of light striking a circular air bubble is illustrated in Fig. 17.2. Rays incident within the sector A–B do not enter the bubble, but are totally reflected from the surface; those between B and C will suffer a number of internal reflections before emerging in an unspecified direction; those between C and D will pass through the bubble, with a slight change of direction. But pores are not normally spherical, and solid grains may have many different shapes, so the result is scattering of light in all directions. However, it still turns out that the scattering effect is roughly proportional to the volume concentration of the inclusion, and is also dependent on size. At very small sizes, it increases rapidly with size, reaches a maximum in a size range comparable with the wavelength of the radiation, and diminishes to a constant value at very large sizes. It is fortuitous that the size of many whiteware body materials, and of the bubbles in glazes, is of the same order as the wavelength of light, so that scattering is a very marked phenomenon in this context.

Lastly, we need to note the phenomenon of absorption. When a ray of light is passing through a dense medium, it suffers a progressive loss of intensity with increasing distance, there being an exponential relation between intensity and thickness.

Thus $I = I_0 e^{-\beta x}$, where $\beta$ is the absorption coefficient. The loss of intensity over a given distance is given by

$$\log_e I_0 - \log_e I = \beta x.$$

The practical significance of this is that when scattering occurs, there is a loss of intensity in passing through the medium from one particle or pore to another. If the matrix itself has a high absorption coefficient, as in the case of body materials, only a small fraction of light will be transmitted. The significance of this will be examined further when we consider translucency.

## 17.2 GLOSS

Gloss is a very important visual feature of glazed ware, especially in the high quality product range. A smooth surface, brightness, and reflection of highlights, add to aesthetic appeal. But gloss is not easy to define or measure. We look now at the property in more detail, and note some of the factors on which it depends.

There are many elements in the subjective assessment of gloss. Among these are the intensity of the specular reflection; the contrast between the specularly reflected light and the diffuse background, seen by tilting the article; the sharpness of the mirror image; and the texture of the surface. No two observers give these factors the same weight, so it is desirable to seek a method of objective measurement. It is usual to work with an incident angle of 45° or 60° and to measure the reflected intensity over a range of angles, which can be done with a simple spectrometer arrangement. The results can be plotted as a polar curve. In the two extremes the reflection would be concentrated in a narrow pencil from a perfectly smooth surface, and spread out in a circle from a perfectly diffusing surface. In practice, something between these two will occur. Some of the incident light will be specularly reflected; some will enter the glaze and be scattered by inclusions or bubbles; some will be reflected from the body:glaze interface; in addition to the emergent scattered light, there will be a degree of diffusion resulting from any irregularity in either the glaze or the body surface. Dinsdale and Malkin (1955) made these measurements on some typical glaze surfaces, with the result shown in Fig. 17.3. The angel of incidence was 60°, and the log of the reflected intensity is indicated by the radial distance from the origin. The black tile glaze, B, shows a very high sharply peaked specular reflection, with very little other reflected light, since all the light entering the glaze is absorbed. The china glaze, A, also shows a high specular reflection with a sharp peak, but there is some diffuse reflection coming from the body surface. The sanitary fireclay glaze, C, shows a lower specular reflection and a broader peak,

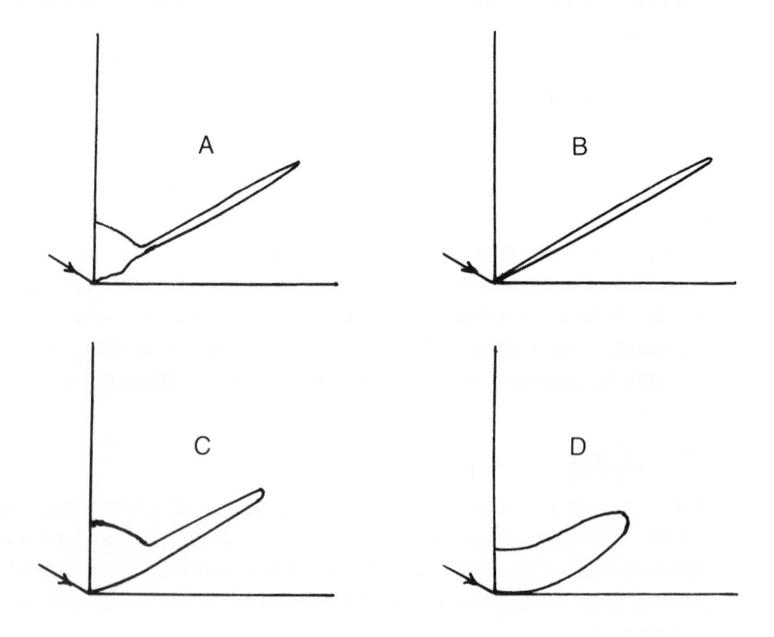


Fig. 17.3 – Polar reflectance curves for four glaze surfaces. A, china; B, black tile; C, sanitary fireclay; D, satin.

due in some measure to the fact that it was opacified, so that there was diffuse reflection from within the glaze itself. The satin glaze, D, shows a much greater broadening of the peak.

Apart from these general observations, the problem is how to quantify these results so as to derive a gloss index. The possible approaches seem to be to use the reflected intensity; the width of the peak; the contrast between the maximum reflected intensity and the diffuse reflection; or the peakiness of the reflection curve. In fact, the best correlation with subjective assessment of gloss was found to be with curve width, which is a measure of the perfection of the reflected image.

It is interesting to speculate on the technical factors that may contribute to good gloss. For example, does the undoubted high gloss of lead-containing glazes derive from the greater fluidity, and hence smoother surface, or from the high refractive index, which may be 1.65 as compared with 1.50 for leadless glazes? The presence of bubble in the glaze is certainly important. Not only does it cause diffuse reflection from inside the glaze, but it leaves minute irregularities in the glaze surface, as was noted in Chapter 13. It is worth repeating the observation by Lepie and Norton, that, for this reason, the gloss is better at the peak firing temperature than it is after the glaze has cooled.

Gloss may diminish appreciably in use as a result of acid attack, attack by alkaline washing agents, abrasion, or deposited films. In fact, change in gloss is one of the most sensitive indicators of structural changes in a glaze surface.

## 17.3 OPACIFICATION

Not all glazes depend on high gloss for their visual appeal. Aesthetic appeal is often achieved by various crystalline effects in the glaze. Opacifiers may be used to cover up undesirable colour in the underlying body, but also to give a solid white appearance that is attractive in its own right. Compositions that allow for a degree of crystallization on cooling may include, for example, oxide of zinc, magnesium, or calcium. If the crystals are large, the so-called crystal glazes are produced. If the crystals are very small, matt or satin surfaces result.

Dense opacification is achieved by the inclusion in the glaze of oxides that do not go into solution during the firing. Among the most effective is tin oxide, which gives a high degree of scattering and a good white. It has been in use for this purpose in glazes for many hundreds of years, being the first opacifier to be discovered. Now it is often partially or wholly replaced by zirconia. This is not as effective as tin and may therefore need to be added in greater concentrations. It may be added as zirconia, or as zircon, $ZrSiO_4$, which dissociates in the glaze, the $SiO_2$ going into solution and the $ZrO_2$ remaining undissolved. Titania is another useful opacifier, although the colour is often somewhat off-white.

These opacification effects are essentially due to the refractive index differences between particle and matrix; it does not matter which of the two is higher, so long as there is a significant difference. This has led to the idea that the same effect can be obtained from scattering at the interface between two immiscible liquids, resulting in the development of so-called emulsion glazes. In these glazes, two immiscible liquid phases develop during the firing, and remain distinct from each other with substantial differences in refractive index, thus producing the opacification effect. An example of this type of glaze is found in the $PbO-B_2O_3-SiO_2$ system. Opacified glazes do not always have a matt surface, but can be produced with a fairly high degree of gloss.

## 17.4 TRANSLUCENCY

Translucency is a well-recognized property of bone china and many porcelains. When a plate is held up to the light, an indistinct image can be seen through it. There is transmission of a small amount, a few per cent only, of the incident light after it has undergone much scattering, so that a sharp image is not transmitted. So translucent occupies an intermediate position between transparent and opaque. It is also often stated that the back reflected light enhances the visual impact of on-glaze decoration. Although translucency may be described in these terms, it is difficult to define precisely as a physical quantity. We take here the simple definition as the ratio of the intensities of the emergent and incident beams. For practical purposes, this ratio is rarely expressed in absolute terms, although the British Standard Specification for china does require a minimum value of 0.7 per cent at a thickness of 2 mm. Industry is usually only interested in relative values quoted in relation to a standard specimen at a given thickness, and their dependence on such factors as body composition and firing conditions.

No simple rules are available. Optimum body compositions with regard to translucency vary with the firing temperature. It can also be said with certainty

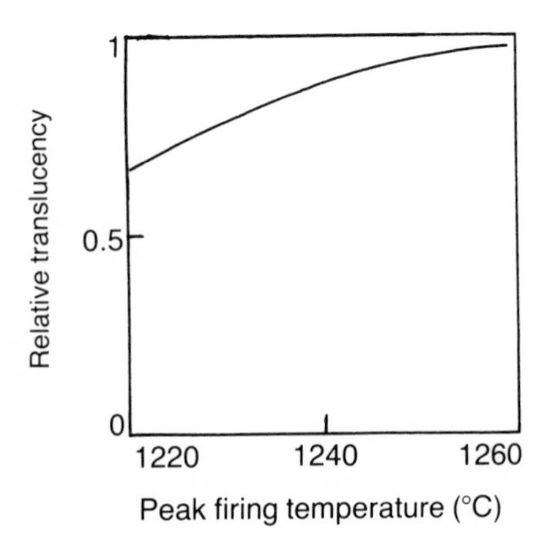


Fig. 17.4 – Effect of firing temperature on translucency of bone china.

that it is impossible to optimize all the required properties at one time. Translucency is very sensitive to temperature, as may be seen from Fig. 17.4. A noteworthy feature is that although vitrification may be regarded as complete when maximum bulk density has been achieved, translucency continues to increase over a period of time if the temperature is held at the peak. So we need to look more closely at the physical processes involved in the body with regard to the transmission of light.

We know that the fired body consists of a glassy matrix, together with crystalline inclusions, and a network of pores. Light is scattered at the surfaces of the particles and the pores, and the loss of intensity depends on the size and the refractive index differential. In between, the light travels through a large number of short paths, in the course of which there will be a degree of absorption. The absorption factor is manifest in the fact that translucency varies with thickness in an exponential manner, as indicated in Section 17.1. Quoted values must always be referred to a standard thickness.

Considering now the effect of refractive index, many experiments have been carried out on model complex systems. In some, crystals of various refractive indices have been immersed in different liquids. In others, the pores of a solid substance have been impregnated with different liquids. Whilst the results show interesting variations the main conclusion is always the same. It is found that the closer the refractive indices of matrix and inclusion are to each other, the greater is the translucency, as shown in Fig. 17.5. In the practical case, it is not possible to achieve a perfect match. For example, in the case of porcelain, the main crystalline components are quartz ($n$ = 1.55) and mullite ($n$ = 1.64) in a matrix ($n$ = 1.50). In bone china, the crystalline phases are anorthite ($n$ = 1.58) and tri-calcium phosphate ($n$ = 1.62), in a matrix ($n$ = 1.50). So the optical

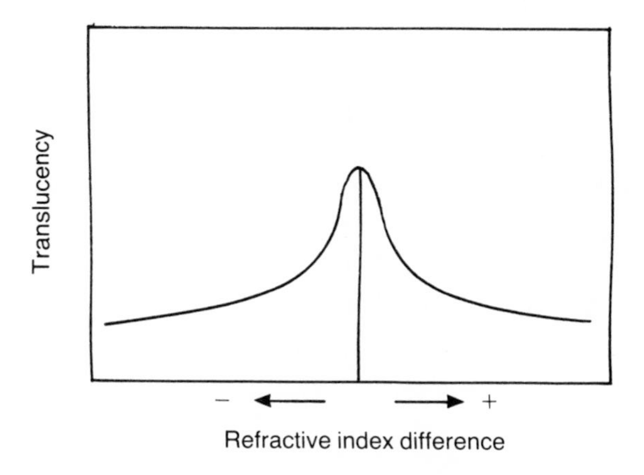


Fig. 17.5 – Translucency of a complex substance.

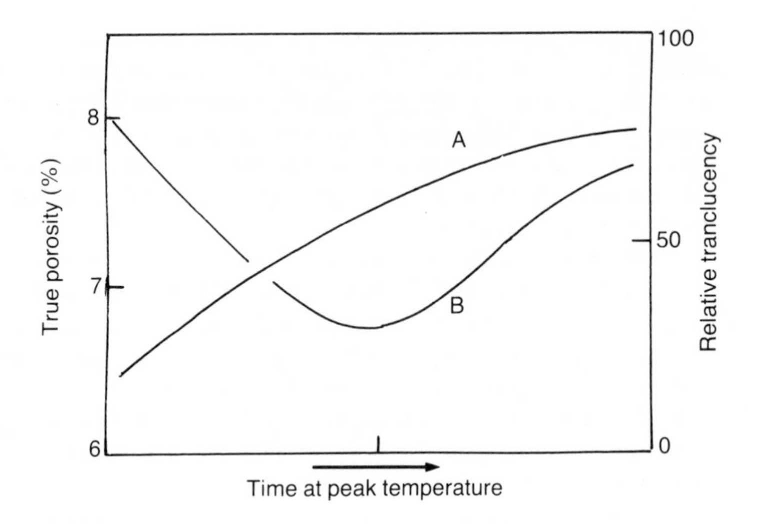


Fig. 17.6 – Relationship between translucency and true porosity. A, translucency; B, porosity.

mismatch is unavoidable and there must always be a lot of light lost in this way. The quantity of glass is also important, and it is probably this factor that varies appreciably with the time at the peak firing temperature.

Scattering by pores is equally significant, and we have already noted that even vitrified bodies have residual closed pores, amounting to as much as 5 per cent volume concentration. The change in pore size distribution at peak firing temperature is the probable explanation for the continuing increase in translucency after maximum density has been achieved. Fig. 17.6 shows how the true porosity of bone china increases with time at the peak temperature after passing through a minimum, whereas the translucency continues to increase with time, even when the porosity is increasing. The reason is that as many small pores disappear, the porosity is made up of much larger pores which, although at increased pore volume, have much less effect on the light absorption than small pores (Dinsdale, 1967).

## 17.5 TOTAL VISUAL IMPACT

Dinsdale (1976) has suggested that the traditional concept of translucency is inadequate as a means of representing the aesthetic appeal of a china article, and that a more comprehensive analysis of the optical properties is required, taking into account absorption and reflection as well as transmission. We can consider all these together as shown in Fig. 17.7. Measurements on a piece of bone china, ground to different thicknesses, are fairly easy to make so as to obtain values of the reflected and transmitted intensities. The absorption can

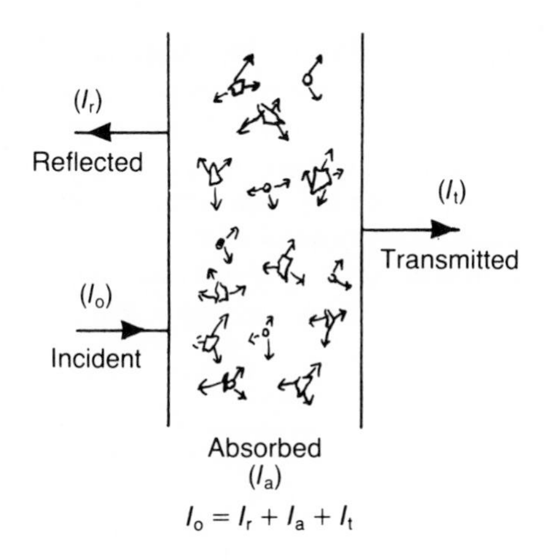


Fig. 17.7 – Passage of light through a ceramic body.

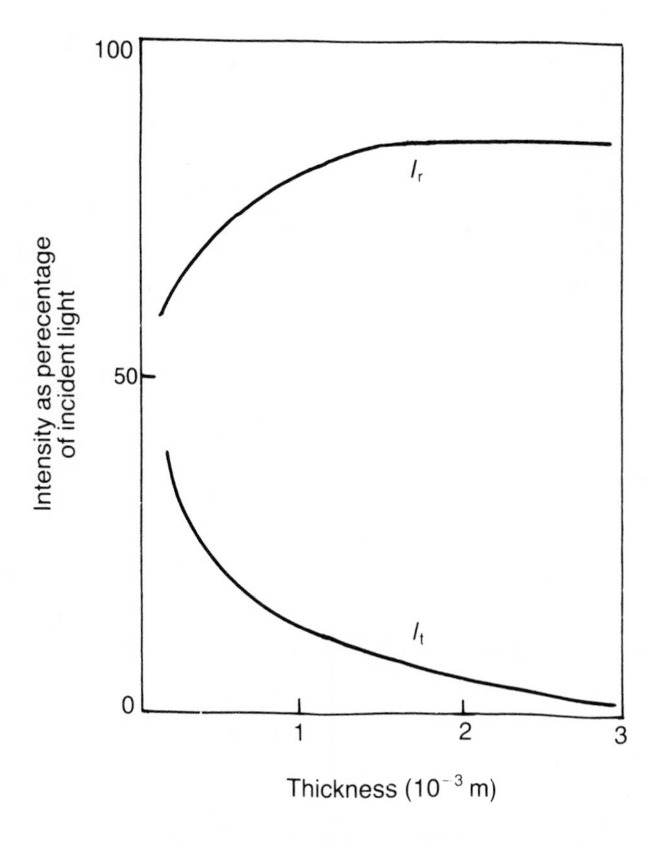


Fig. 17.8 – Transmission and reflection for a sample of bone china at various thicknesses.

then be deduced by difference. The results of such measurements on bone china are shown in Fig. 17.8. The transmitted intensity increases in an exponential manner as the thickness is reduced, and must come to 100 per cent at zero thickness. The reflected intensity becomes fairly constant at practical thicknesses, and accounts for a very large proportion of the light, both the absorption and the transmission being relatively small.

Formal analysis of the reflection and transmission data can be carried out following the Kubelka-Munk basis for turbid media, and it is possible to derive

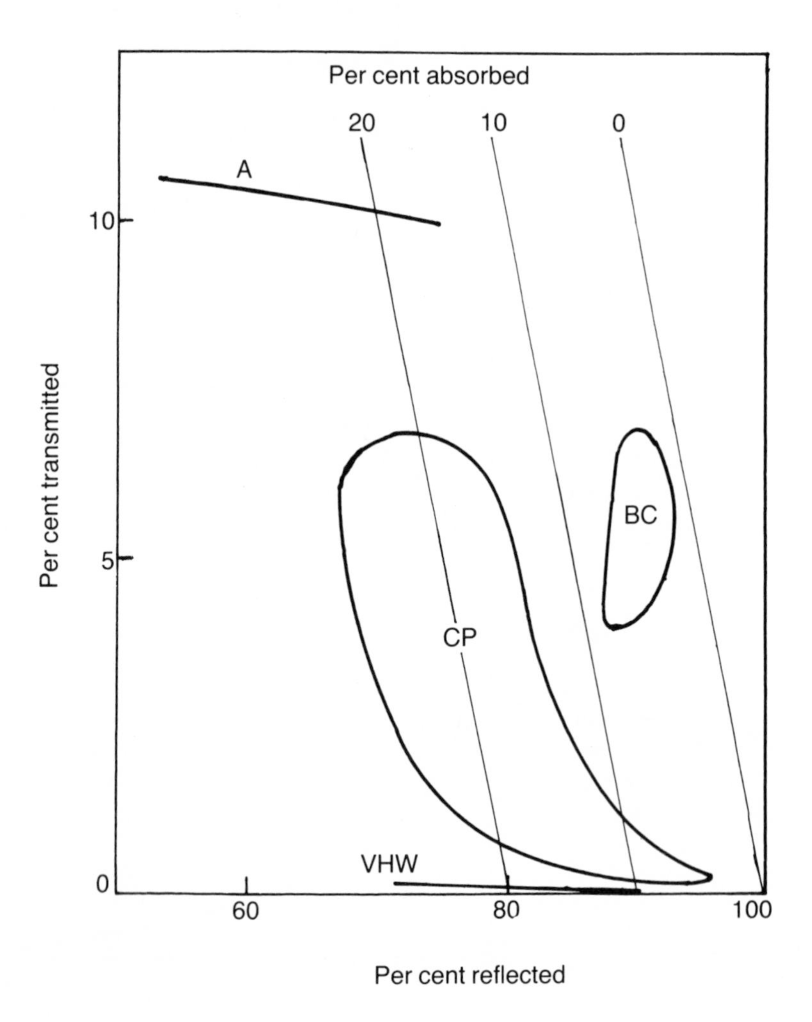


Fig. 17.9 – Reflection, transmission, and absorption for commercial bodies at a thickness of $2 \times 10^{-3}$ m. A, American porcelain; BC, bone china; CP, continential porcelain; VHW, vitrified hotel ware.

scattering and absorption coefficients that are characteristic of the material and independent of the thickness. However, the same general conclusions can be arrived at by using the reflection and transmission values themselves, at a standard thickness, as shown in Fig. 17.9. By difference from 100 per cent the absorption values can be obtained, and plotted on the diagram as a fairly straight line. Measurements have been made on a random sample of commercial bodies, and the results are shown in Fig. 17.9. For thirteen bone china samples, made in different parts of the world, the values fell within a fairly narrow area. Forty samples of continental porcelain gave values within the area shown, which shows a much wider spread. Values for three American porcelains, and for six British vitrified hotel ware samples, are also shown. For comparison, it is of interest to note that a piece of opal glass had the coordinates 60, 40, and a piece of Perspex 67, 30.

We may now consider the broad trend of these results. The vitrified hotel ware has a very low translucency, and does not claim to be translucent. The bone china results represent high translucency, but also high reflectance, and hence low absorption, below 10 per cent. It is suggested that it is the high reflectance, as much as the high translucency, that gives bone china its distinctive appearance. It is very white because of the low absorption. Continental porcelains show a wider range of transmission but an appreciably higher absorption. The American results are interesting in that they show higher transmissions than bone china, but a much lower reflectance and much higher absorption. They thus have a quite different visual impact. Their values are characteristic of porcelains that have an added glass component in the body.

It is not suggested here that this treatment is adequate to represent all the visual properties, but that something more than a simple translucency figure is desirable.

## REFERENCES

Dinsdale, A. and Malkin, F. (1955) *Tr. Brit. Cer. Soc.*, **54**, 94.
Dinsdale, A. (1967) *Sc. of Ceramics*, **III**, 323.
Dinsdale, A. (1976) *Amer. Cer. Soc. Bulletin*, **55**(11), 993.

# 18

# Properties of finished ware — electrical

The early development of electrical ceramics was stimulated by the increasing use of electricity for heating, lighting and power, followed by communications. When organic materials proved to be inadequate in hostile environments, with particular reference to temperature, moisture and pollution, the high durability of porcelain came to be recognized. Porcelain components are widely used for support insulators, switch components, and many other ancillary functions. As voltages have increased, now often up to 400 kV or even higher, insulators have become larger and strength has acquired increased significance. The main disadvantages of porcelain are that its dielectric strength decreases with higher temperatures, and that it is not suitable from a loss point of view at high frequencies. Other types of ceramic body have been developed for special purposes. For example, there are low thermal expansion bodies, like cordierite, for conditions of high thermal shock; titania based bodies for high dielectric constant; and steatite for high frequency, low loss, uses.

Recent developments in electronics have highlighted the importance of electrical properties, and their dependence on such factors as composition, moisture, temperature, frequency, and so on. There is now a bewildering range of options open to the electrical enginer. Bloor (1970) and Johnson and Robinson (1975) have reviewed body compositions and modern manufacturing techniques. In addition, Bloor (1953) has compiled a very comprehensive review of the composition and properties of electrical ceramics. Another excellent summary of the properties appropriate for particular uses is to be found in Kingery *et al.* (1976).

## 18.1 ELECTRICAL THEORY AND DEFINITIONS

If a steady potential difference, $V$ volts, is applied across a specimen with resistance $R$ ohms, a steady direct current $j$ amperes will flow, such that $j = V/R$. If the specimen has a length $l$ and a cross-sectional area $A$, then $R = \rho(l/A)$, where $\rho$ is a constant known as the resistivity. It is the resistance in ohms of a unit volume of the material, and may be expressed in the unit ohm-m. Its reciprocal, $\sigma$, is the electrical conductivity, expressed as mho.m$^{-1}$.

In many electrical applications the applied potential is not direct, but alternating, in sinusoidal manner $V = V_0 \sin \omega t$. The frequency of this alternating voltage is $\omega/2$ cycles per second, generally expressed as Hertz (Hz). If this voltage is applied to a circuit containing a resistance $R$, a capacitance $C$, and an inductance $L$, the current will also vary in a sinusoidal manner, but will differ in phase from the voltage. The maximum value of the current $j_m$ will be given by

$$j_m = \frac{1}{\sqrt{2}} \frac{V_m}{\sqrt{R^2 + \left(\omega L - \dfrac{1}{\omega C}\right)^2}}$$

where $V_m$ is the peak voltage during the cycle. The effective current $j_e = j_m/\sqrt{2}$.

The angle by which the current lags behind the voltage is usually denoted by $\delta$, and

$$\tan \delta = \frac{\left(\omega L - \dfrac{1}{\omega C}\right)}{R}$$

The expression in brackets is known as the reactance, and consists of an inductive reactance, $\omega L$, and a capacitative reactance $1/\omega C$. The quantity which corresponds to the resistance under d.c. conditions is the impedance, $Z$, and

$$Z = \left\{ R^2 + \left(\omega L - \frac{1}{\omega C}\right)^2 \right\}^{\frac{1}{2}}$$

Apart from insulation, materials may also be used for their dielectric properties in condensers and similar appliances. If an electric charge $Q$ is stored in a material with capacitance $C$ under a voltage $V$, then $Q = C \times V$. For a parallel plate condenser, the capacitance

$$C = \frac{\text{Area}}{\text{thickness}} \times \text{E}$$

where E is the permittivity or dielectric constant. Under alternating conditions, a material with a loss angle $\delta$ and a permittivity E will suffer an energy loss E $\times$ tan $\delta$. For many purposes it is desirable to keep this loss as low as possible, but if high values of E are required, then very low values of tan $\delta$ must be sought.

At very high voltages, insulating materials may break down, and a sudden surge of current may occur, damaging the component and rendering it useless. The voltage at which this occurs is known as the dielectric strength or puncture strength. Its value may be as low as 100 V $mm^{-1}$ for some ceramics, but for vitreous porcelains it is more likely to be in the range 20-40 kV $mm^{-1}$. Its value is highly dependent on the method of measurement.

So over the whole range of electrical uses, the material properties that are of significance include resistivity, permittivity, loss, and dielectric strength. We consider now what values these properties may have for whiteware ceramics, and how they may vary with such factors as time, temperature, frequency, porosity, composition, and moisture.

## 18.2 RESISTIVITY

Solids can be divided into three main categories with respect to the facility with which they conduct electricity. In metals the conduction is by means of electrons, which are freely available, and the resistivity is thus very low. At room temperature it is generally in the range $10^{-8}$ to $10^{-6}$ ohm-m, increasing with temperature. For copper, for example, it is about $1.5 \times 10^{-8}$ ohm-m. Semiconductors have intermediate values between metals and insulators; silicon carbide has $\rho = 10^{-1}$ ohm-m approximately. In ceramic insulating materials there is very little electronic conduction; the current is mainly carried by mobile ions, such as sodium and potassium. The resistivity of electrical porcelain is in the range $10^{11}$ to $10^{12}$ ohm-m at room temperature.

### 18.2.1 Composition

High tension porcelains are generally to be found in the triaxial system clay:quartz:felspar, fired to around 1250° C to a high degree of vitrification. Pegmatites are often used as the flux instead of pure felspar. The fired body consists of an alkali-silicate glass with crystals of quartz and mullite. In this complex system, the crystals and the closed pores have a high resistivity, and most of the conduction takes place in the glass. The conduction is mainly ionic. In particular, the proportions of soda and potassium are of great importance. Budnikov and Maslennikova (1966) studied the influence of the composition on the properties, and showed that potassium felspars gave much higher resistivities than sodium. They suggested that in order to achieve adequate resistivity values, e.g. $10^{11}$ ohm-m at room temperature, the $Na_2O$ content should not be greater than 1 per cent. They also showed that replacement of these alkali oxides by oxides of Mg, Ca, or Ba improved the resistivity. Some workers have found a roughly linear relationship between resistivity and $K_2O$:$Na_2O$ ratio. This ratio also influences the $T_e$ value, (the temperature at which $\rho = 10^4$ ohm-m), increasing it by 50 to 100° C with increasing potassium. The increased conduction associated with sodium is because it has a much smaller ionic radius than potassium.

The other compositional feature that is worthy of note is the special characteristics of magnesium-silicate bodies. MgO itself is a very good insulator. Bodies based on high talc contents, up to 90 per cent, can have resistivities higher than porcelain by a factor of 10. However, the crystalline phases developing in magnesia-containing bodies are complex and not easy to control. An interesting example is discussed by Bloor (1964).

High alumina bodies and zircon porcelains can also have resistivities as high as $10^3$ ohm-m.

### 18.2.2 Time

When an electric potential is applied to an insulator there is a large rapid initial current, known as the charging current, which then decays steadily over a period of up to one minute at room temperature, this being known as the absorption current. Following this there is a smaller steady current, associated with the true value of the resistivity.

Over long periods of time, particularly under conditions of d.c. stress, it might be expected that polarization would occur owing to the continued transport of ions to the anode. In fact, this results in a slight increase in resistivity, but the difference is of no significance under normal working conditions.

### 18.2.3 Temperature

For some applications and conditions, insulators may require to maintain adequate resistivity to temperatures appreciably above atmospheric. It is thus of interest to see how resistivity varies with temperature. Measurements over a range of temperatures show a decrease in resistivity as the temperature is increased. Above a transition temperature, which may be around 100° C, there is an accelerated increase in conduction, due to thermal effects. There is thus a discontinuity at this point.

A number of formulae have been suggested for the relationship between resistivity and temperature, but the experimental data are well fitted by the well known Rasch-Hinrichsen equation

$$\rho = A\ e^{B/T} \quad \text{or} \quad \log \rho = \log A + B/T$$

where $T$ is the absolute temperature, and $B$ is some kind of activation energy. A plot of log $\rho$ against $T$ would give a curve such as that shown in Fig. 18.1(A). If log $\rho$ is plotted against $1/T$, a straight line is obtained, as in Fig. 18.1(B). In practice, for a typical porcelain, the facts are better represented by the dotted lines, showing the discontinuity at the transition temperature.

### 18.2.4 Porosity

Because of the high strength requirement, porcelains are fired to a high degree of

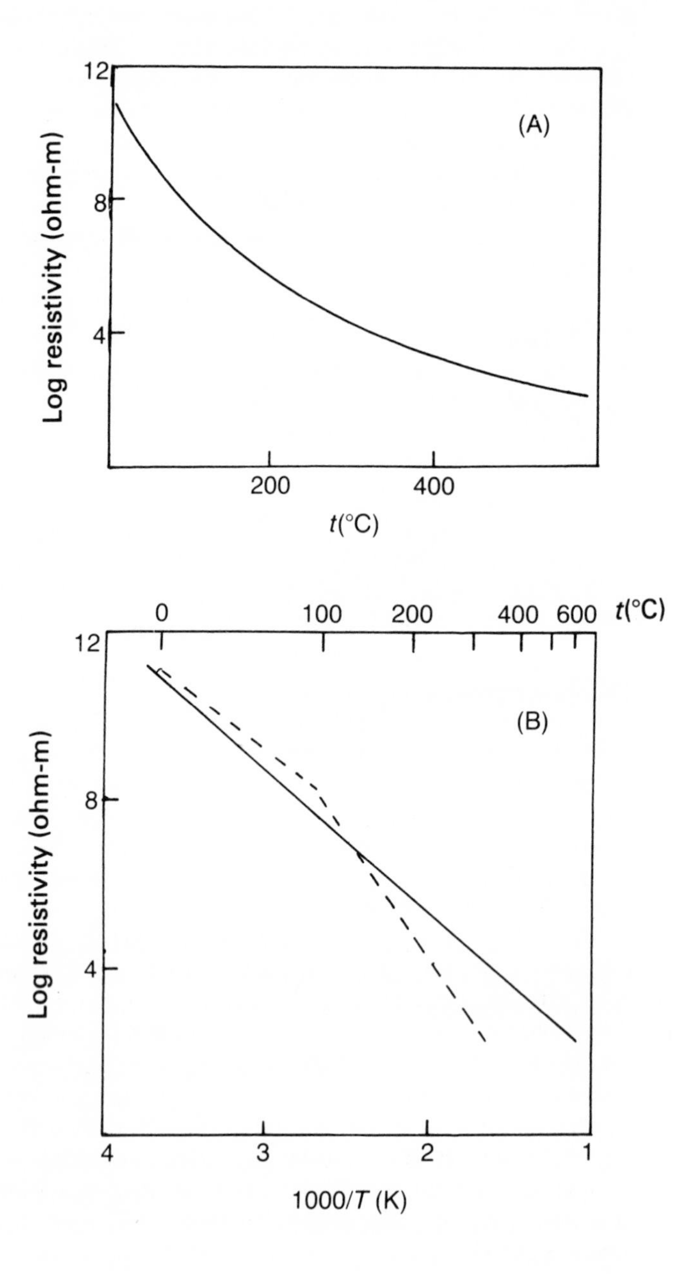


Fig. 18.1 – Variation of resistivity with temperature for a typical high-tension porcelain. —— Rasch-Hinrichsen, – – – typical porcelain.

vitrification. However, there is always a residual proportion of closed pores amounting to 5 to 8 per cent by volume. Pores are poor conductors of electricity, so the influence of porosity over a limited range is slightly to increase the resistivity, roughly in proportion to the pore volume. Over wider ranges of porosity the effect is much more complex than this.

### 18.2.5 Moisture

Insulators are sensitive to atmospheric moisture, especially in regard to the surface conduction. It is presumably due to the presence of a conducting film of water on the surface. The surface resistivity may fall by one or two orders of magnitude under conditions of high relative humidity. The presence of accessible porosity, of course, could lead to an unacceptable loss of resistivity due to the penetration of water.

### 18.2.6 Frequency

Impedance is dependent on frequency, but at most working frequencies, impedance and resistivity are not very different. At room temperature, the dielectric conduction increases with frequency, and so the resistivity decreases with frequency. The d.c. resistivity is somewhat higher than the impedance at room temperature. As the temperature increases, the values tend to come together as the effect of frequency diminishes.

## 18.3 PUNCTURE STRENGTH

Puncture strength, or dielectric strength, is the voltage at which a material fails owing to a sudden drop in resistivity. The high surge of current that occurs is sufficient to produce physical damage within the material and so render it unfit for use.

The value is usually quoted in volts per mm. Results show very large variations, mainly due to the influence of the method of test, and shape and thickness of the test specimen. The thickness of the specimen affects the rate of heat dissipation; thin specimens do not undergo as much temperature rise as thick ones, and thus give a higher value. The time for which the stress is applied, and the manner of stress increment also influence the result. Vitreous porcelains usually have a dielectric strength between 20 and 40 kV $mm^{-1}$. There is evidence that for extruded specimens the dielectric strength is higher in the direction parallel to extrusion than it is in the perpendicular direction. Values for a.c. and d.c. are in fairly good agreement, provided that the comparison is with the peak a.c. voltage, and not the root mean square. Body composition is not of major importance, but increased porosity can result in a marked lowering of dielectric strength.

The effect of temperature is in some ways similar to that noted in connection with resistivity, in that it may be represented by two intersecting straight lines. At low temperatures, up to about 100° C, the puncture strength is fairly constant. When thermal effects begin to operate, at temperature between 100 and 200° C,

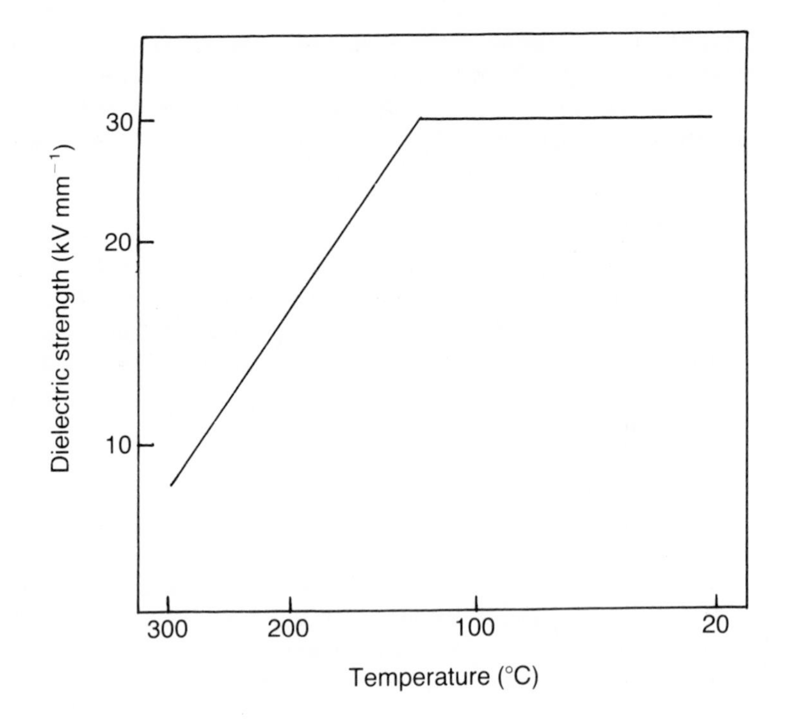


Fig. 18.2 – Variation of dielectric strength of electrical porcelain with temperature.

there is a transition to a rapid drop in strength with increasing temperature. The value of the transition temperature is again very much dependent on the test conditions. It tends to be higher for potassium than for sodium based bodies. A typical plot for porcelain is shown in Fig. 18.2. In practice, the dielectric strength of porcelain insulators is well in excess of requirements.

## 18.4 PERMITTIVITY AND DIELECTRIC LOSS

In applications where capacitance is of importance, high values of the dielectric constant, or permittivity, may be required. For most ceramic materials the value lies between 5 and 8, zircon and alumina porcelains being towards the top of this range. For porcelain, the value is not greatly dependent on composition. Outside the range of triaxial porcelains, however, are the titania based materials, which have outstanding properties. $TiO_2$ crystals have permittivities up to 100, and alkali earth titanates can have values up to $10^4$. These high values have assisted in the process of miniaturization, giving high capacity to size ratios. Permittivity of porcelain falls slightly with increasing frequency, and increases slightly with

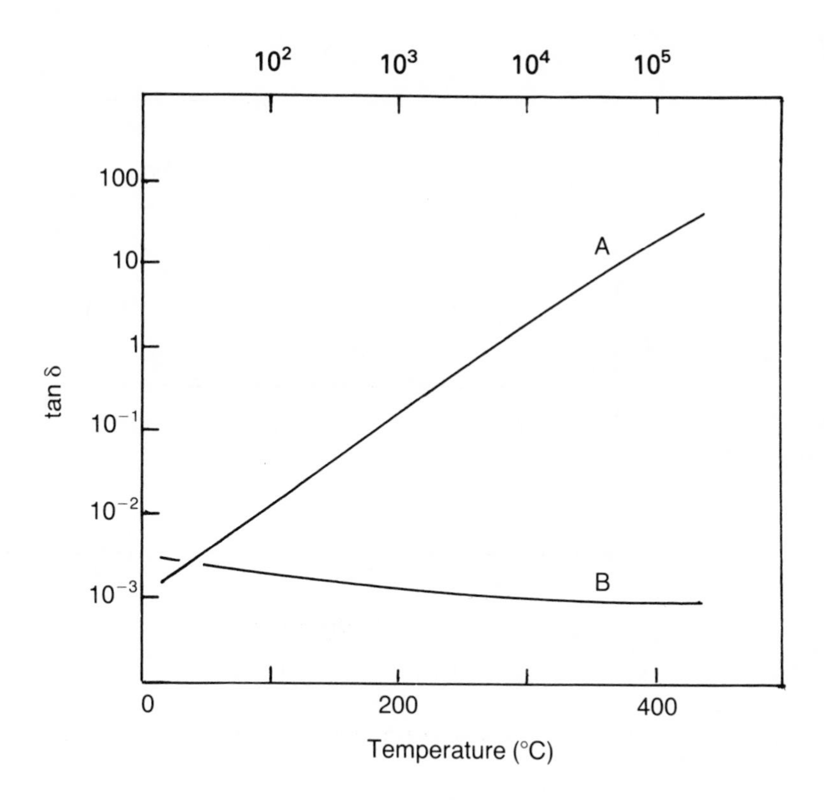


Fig. 18.3 – Variation of tan δ with temperature and frequency for typical porcelain. A, variation with temperature at $10^3$ Hz; B, variation with frequency at 20° C.

increasing temperature, but neither rate of change is large. The value decreases with increased porosity.

More important, perhaps, for many high-frequency applications is the dielectric loss, which depends on both the permittivity and the loss angle tan $\delta$. Porcelain has a relatively high loss and is not suitable for radio frequency use. Loss is mainly associated with alkali ions, so in compositions low in soda or potassium much lower losses can be achieved, for example, in steatite, alumina, and titanate bodies. At room temperature the value, for 50 Hz, may be of the order of $100 \times 10^{-4}$ for porcelain, and as low as $10^{-4}$ for low-loss bodies. Tan $\delta$ varies with both frequency and temperature. For porcelain tan $\delta$ increases rapidly with temperature, especially at low frequencies; at room temperature it falls slightly with increasing frequency, as shown in Fig. 18.3.

Kingery *et al.* (1976) has suggested that the frequency and temperature effects are not independent, and are best plotted together on the same graph, in the manner shown in Fig. 18.4. This treatment emphasizes the fact that the effect of frequency is more evident at higher temperatures, and that the effect of temperature diminishes at high frequencies.

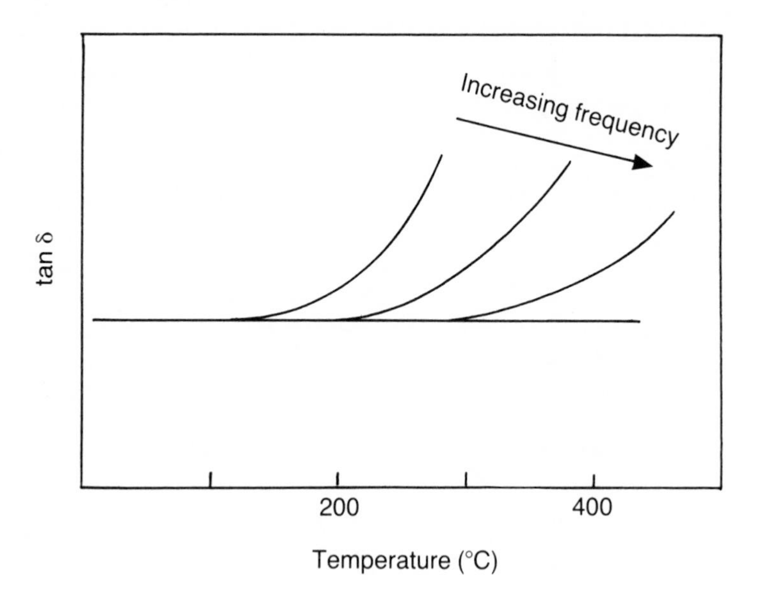


Fig. 18.4 – Variation of tan δ with temperature at various frequencies.

## 18.5 LOW THERMAL EXPANSION BODIES

In conditions which present an appreciable element of thermal shock there is need for a body with good thermal shock resistance combined with sufficiently high resistivity. Electric fire bars, support elements in cookers and heaters, and cores for electrical resistors are common examples. Dimensional stability may also be important even at high frequencies. This requirement is well met by cordierite bodies. Commercial cordierite bodies cover a wide range of composition, but they all depend for their function on a substantial cordierite content in the fired state. Cordierite, $2MgO.2Al_2O_3.5SiO_2$ is formed at high temperatures when clay is mixed with a magnesia-bearing mineral such as talc, and has a very low thermal expansion. The expansion coefficient may be as low as $2.5 \times 10^{-6}$ compared with about $6 \times 10^{-6}$ for porcelain.

Cordierite bodies have a room temperature resistivity of about $10^6$ ohm-m for the porous version, and about $10^{10}$ to $10^{12}$ ohm-m when vitrified. The resistivity falls with increasing temperature, by about two orders of magnitude to 300° C.

Other low expansion bodies can be found in the lithium aluminium silicate system.

## 18.6 CONDUCTING GLAZES

One of the problems with high tension insulators is the difficulty of achieving uniform field distribution. The uneven field resulting from geometrical considerations may be exacerbated by atmospheric pollution, especially in conditions of high humidity. In severe circumstances, localized discharges may occur, giving rise to radio interference, and in extreme cases flashover and failure of the insulator. One means of reducing this danger is to provide a limited current leakage by the application of a semiconducting surface layer. This can be done by incorporating a conducting phase in the glaze.

Clarke *et al.* (1961) described the use of $TiO_2$ for this purpose, and Binns (1971, 1974) has reviewed the use of other oxides, coming to the conclusion that glazes containing $SnO_2$ with a small addition of $Sb_2O_5$ gave the most promising results. In addition to the right level of electrical resistance, they gave good corrosion resistance and good thermal expansion characteristics so as to maintain mechanical strength. It was also found that the concentration of conducting phase needed to be just that required to provide a continuous conducting network of particles, and was in the region of 10 per cent.

The surface resistivity was in the region of $10^7$ ohms per square at room temperature, falling slightly to around $6 \times 10^6$ ohms per square at 350° C. For a glaze thickness of 0.2 mm, these values correspond to volume resistivities of 5 and $3 \times 10^6$ ohm-m, respectively.

A rather similar problem arises in the case of floor tiles in operating theatres, where electrostatic discharges in the presence of volatile gases can be dangerous. The risks can be eliminated by the use of a conducting phase in the tiles, giving a slight lowering of their resistivity.

## REFERENCES


Binns, D. B. (1971) *Tr. Brit. Cer. Soc.*, **70**, 253.

Binns, D. B. (1974) *Tr. & J. Brit. Cer. Soc.*, **73**(1), 7.

Bloor, E. C. (1953) *Ceramics – A Symposium* (British Ceramic Society), p. 227.

Bloor, E. C. (1964) *J. Brit. Cer. Soc.*, **1**(2), 309.

Bloor, E. C. (1970) *J. Brit. Cer. Soc.*, **7**, 77.

Budnikov, P. P. and Maslennikova, G. N. (1966) *International Ceramic Congress*, 155.

Clarke, C. H. W., Turner, R. B., and Powell, D. G. (1961) *Tr. Brit. Cer. Soc.*, **60**, 330.

Johnson, P. and Robinson, W. G. (1975) *Tr. & J. Brit. Cer. Soc.*, **74**(4), 147.

Kingery, W. D., Bowen, H. K. and Uhlmann, D. R. (1976) *Introduction to Ceramics*, 2nd ed., 951–955.

# 19

# Whiteware products in use

We have already touched on some of the hazards encountered by whiteware products in use, including mechanical and thermal shock. We look now in more detail at the general problem of fitness for function and durability in service. Ceramics have a high survival potential, in the sense that they can remain unaltered in essential properties, over thousands of years, but in their normal daily use there are aggressive aspects of the environment that need to be taken into account.

## 19.1 MOISTURE EXPANSION

It is well known that many materials change size in the presence of water, in either the liquid or vapour phase. Some, like string, contract; others, like wood, expand. Porous ceramics expand progressively over long periods of time; to the extent of approximately 0.1 per cent linear increase. One practical consequence of this is that glazed articles may suffer crazing, that is a network of fine cracks, making the article unacceptable. This only happens if the glaze compression is inadequate; in fact, so much is known about the factors involved that crazing is now very much a thing of the past. It is important to note that moisture expansion can only occur in porous bodies, so all vitreous ware is immune from it, except in rare cases where a finite degree of porosity in the surface layer can cause trouble. The other important aspect of it is that it may cause trouble with the adhesion of wall tiles, though here again modern fixing techniques and materials have reduced the problem to negligible proportions.

### 19.1.1 Some common features

Although the extent of moisture expansion may vary with the nature of the material, there are some well established features that are common to all. These have been described in varying degrees of detail by Norris *et al.* (1958) and Dinsdale and Vaughan (1962). The most notable feature is that moisture expansion begins as soon as the material leaves the kiln, proceeds very rapidly in the early stages, but may continue at a very slow rate over a period of a year or two. A typical curve of moisture expansion against time is shown in Fig. 19.1. The curve is of the exponential type, but is different in shape for different materials. The eventual equilibrium expansion is usually of the order of 0.1 per cent for whiteware bodies. The expansion is higher for higher porosities, but in fact is more directly related to internal surface area, which in the case of these bodies is usually of the order of 1 $m^2$ $g^{-1}$. Increased humidity, temperature, or pressure accelerate the expansion process.

### 19.1.2 The underlying process

It is now well established that the expansion is caused by the adsorption of water vapour, which reduces the surface energy and so results in volume expansion. If a body is completely desorbed by heating to a high temperature, it is

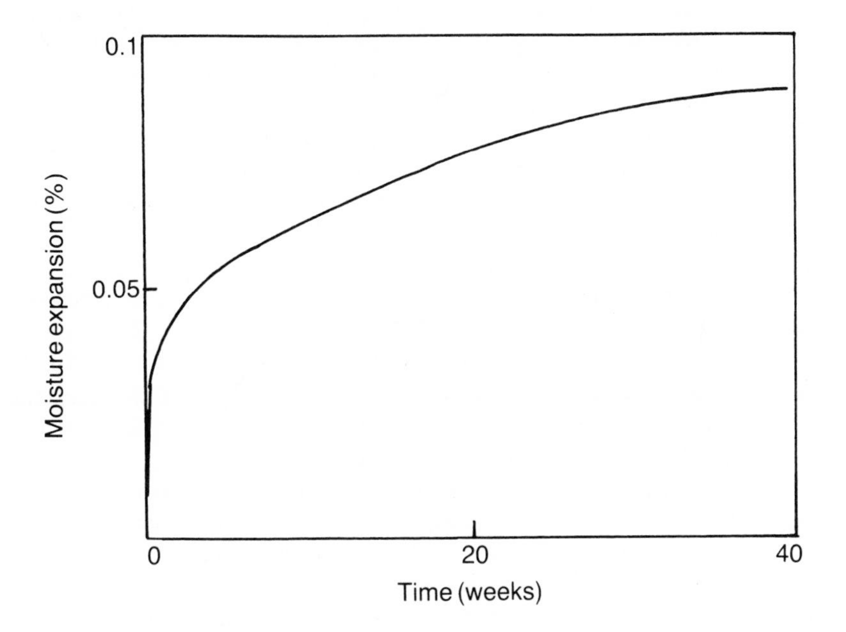


Fig. 19.1 – Moisture expansion against time for a porous whiteware body.

found that it readsorbs moisture immediately on cooling. The first stage is a physical adsorption in which one or two molecular layers are involved, sufficient to give a rapid initial expansion. Following this, the body continues to take up water over a long period of time, and there is a continuing positive, though small, expansion. This expansion can only be reversed, and the water driven off, by heating to about 1000° C.

It is found that the penetration is mainly into the glassy matrix of the body; crystalline components show very little expansion. As the water penetrates the glass it breaks some of the network-forming bonds and loosens the structure. The penetration of water has been experimentally verified by Moulson and Roberts (1960). Body factors that influence the degree of expansion include porosity, internal surface area, and composition and amount of glass. The relationship with porosity is complex, because of the changing pore size distribution with firing temperature, but the connection with surface area is clearcut.

The other important feature is that the expansion can be accelerated by an increase in temperature or pressure. Harrison and Dinsdale (1964) have shown how these two parameters can be separated experimentally, and their relative importance assessed. They showed that temperature was of prime importance.

### 19.1.3 Glazed articles

Most of the work on moisture expansion has been carried out on unglazed bodies. For articles such as wall-tiles, exposed on one side, the relevance of this is clear. In the case of completely glazed articles, such as tableware, access is not so obvious. However, there is often a limited degree of access through stilt marks or unglazed feet. In any case, moisture expansion is not dependent on the presence of liquid water. Humidity in the surrounding atmosphere is reflected in the pores after a period of time, since glazes are not impermeable to vapour. The general characteristics of moisture expansion apply equally to glazed ware, though there may be a shift in the time scale as compared with unglazed material.

### 19.1.4 Some practical aspects

Testing for moisture expansion and crazing has traditionally been based on the use of high pressure steam, but, as with many other accelerated tests, this procedure is open to some objection. It has been shown that the expansion associated with a given weight of water adsorbed is very different at room temperature as compared with high temperature steam. The changes taking place in the body are different. In addition to this, high temperatures in the test may reduce the glaze compression in cristobalite-containing bodies in a manner that is in no way related to room temperature conditions. Thus, for both these reasons, results with high pressure steam should be treated with caution.

So far as wall tiles are concerned, the use of modern adhesives, flexible where necessary, and expansion joints has virtually eliminated adhesion failures due to moisture expansion. The risk of crazing cannot be completely eliminated, even with high crazing resistance tiles, if these are fixed on unstable backgrounds imposing additional stresses.

On the whole, the improved technological understanding and control has virtually eliminated the crazing problem.

## 19.2 CHEMICAL DURABILITY

Domestic tableware in use comes into regular contact with both acid and alkaline substances, and the glaze and decoration surfaces are thus liable to suffer chemical attack. Colours are, in general, more vulnerable than glazes, since the glassy fluxes in which they are formed mature at much lower temperatures than do glazes. One way of increasing resistance to chemical attack is to use underglaze colours, but this has the disadvantage of a restricted colour range. The two main areas of possible attack are connected with washing up procedures, and contact with acid foods. We consider these two areas in turn.

### 19.2.1 Washing agents

All tableware used in contact with food requires to be washed; several times a day in the case of domestic use, but very much more frequently in the case of hotel and restaurant use. Invariably, an alkaline washing agent is used, and this may be of such a nature as to attack the surface of the ware after prolonged exposure. Any defects in the glaze surface, or any abrasion or scratching, which increase the exposed surface area, increase the rate of attack. The attack usually begins by reducing the gloss, but in severe cases may result in the removal of the colour.

Franklin and Tindall (1959) studied the effects of different types of washing agent, having first established the fact that a continuous immersion test could be used to simulate actual service conditions involving repeated washings. On the ceramic side, they found that harder decorating firing gave better resistance to attack. With regard to the washing agents, they were not able to find a comprehensive theory for the chemical process, but showed that pH on its own was not a reliable guide, and that complex materials did not always reproduce the characteristics of their individual components. Certain very important practical conclusions did emerge from their work. Concentration of washing agent did not have any appreciable effect within a range on either side of that recommended. Attack increased progressively with time, but most significant was the effect of temperature. Particularly in the range often encountered in dishwashers, the attack increased very rapidly with temperatures above 140° F. The nature of the washing agent was also important. A curious feature was that pure soap flakes and soap powders were found to be potentially very harmful. Modern synthetic detergents, based on surface-active components, were found to be relatively harmless. However, these cannot be used in dishwashing machines, and in this field washing agents based on soda ash, sodium phosphate, and sodium silicate are generally used. The problem here is that temperatures have to be high enough for the ware to dry itself, but beyond this, excessive temperature is very harmful. It was also clear that some commercial washing agents were very much less harmful than others, without any diminution in their cleaning power.

It is thus clear that, in order to minimize the alkali attack on decorated ware, it is first necessary to choose the correct washing agent; a surface active detergent for hand-washing, and an agent for dishwashing machines that shows minimum corrosion potential. Some washing agents include aluminate additions, in which case adsorption of $Al^{3+}$ ions on the surface reduces the attack. In addition, excessive temperatures should be avoided; not more than 140°F in machines. It is not, of course, possible to wash by hand at temperatures much above 110°F, but a bad practice that is sometimes encountered is to place the ware in a bowl, and leave it soaking for some time in boiling water.

Hotel ware washed in industrial washing machines faces a difficult environment. Many washings a day may be needed; and time cycles may be very short, so that a strong cleaning action is needed. For these reasons manufacturers of hotel-ware pay special attention to the production of glazes with high resistance to attack.

### 19.2.2 Acid attack

Tableware and cooking ware are often in contact with acid substances, such as fruit juices, vinegar, and condiments. Prolonged exposure may result in attack on decoration. It is found that some colours are much more vulnerable to acid attack than others. It is also noteworthy that different acids have different effects; oxalic acid is much more corrosive than either acetic or hydrochloric. Attack again increases with time, so that prolonged storage of acidic foods on decorated surfaces is to be avoided.

## 19.3 METAL RELEASE

There has been much debate in the last twenty years about the health hazard said to be associated with the extraction of toxic substances, such as lead and cadmium, from glazed and decorated ware into acid foodstuffs. Anxiety has been generated by a small incidence of poisoning occurring in non-typical circumstances, namely, long storage of food or liquid in vessels glazed with high lead glazes inadequately fired. There is no evidence that any health hazard exists from the use of industrially produced, as distinct from craft made, tableware. However, concern about lead pollution from petrol and other sources has led to pressure for a reduction to the minimum achievable of the contribution of pottery to the general exposure levels. To achieve this a number of national and international standards have been established, using a simulative test based on 4 per cent acetic acid at room temperature. Limits are generally expressed in terms of mg per litre for hollow ware, and mg per square decimetre of exposed surface for flatware.

It is now well established that the quantities extracted from this test are very much higher than those extracted by a wide range of acid foodstuffs or liquids. Moreover, the amount extracted by foodstuffs is very small compared with that already in food, and diminishes rapidly with repeated use. It would seem that the health hazard has been grossly exaggerated, and that the current regulations impose an unnecessarily rigid technological restriction on the pottery industry.

From the manufacturer's point of view, there are certain basic factors that need to be recognized if the product is to meet the regulatory requirements. One is that the unavoidable statistical spread of results makes it necessary to set a production control level considerably lower than the standard limits. This is especially important in cases where the regulations rely on results on a single test piece. Among the ceramic parameters that have a bearing on metal release is that of the enamel kiln firing temperature. Release from on-glaze colours is much higher than for glazes, on account of the lower firing temperature fluxes that have to be used. Metal release can be substantially increased by inadequate enamel firing, in terms of either temperature or time. Composition plays an important part, high silica content being beneficial. It must be noted that colour composition can be strongly influenced by ion migration from the underlying glaze; potassium ions migrate freely and tend to increase metal release. Further, ion migration can take place from the body; there are cases on record where high potassium fluxes in bone china have given rise to significant changes in the decoration layer. Kiln atmosphere has an effect. Water vapour is beneficial, but in some circumstances a condensation film on a glaze from its own vapour can increase metal release.

Notwithstanding the clear influence of flux composition, the nature of the colouring pigment itself is important. Corrosion is sometimes found to be initiated at the pigment particles. For this reason, recent years have seen the elimination of certain colouring oxides, in some cases replaceable but in others not. Overall there is a reduction of palette choice.

The problem of cadmium is difficult in that there is no known replacement in the very important yellow, orange and red area of the spectrum, traditionally based on cadmium-sulfo-selenide pigments. It may be necessary to find a more radical solution here.

Cooking ware presents a special problem. Acid foods may well be held at high temperatures for a long period of time. For this reason most cooking ware is now undecorated on the internal surfaces, and in many cases leadless glazes are used.

## 19.4 ABRASION

The concept of hardness of glazes is not easy to quantify in that it is not a single physical property. It implies resistance to surface damage by scratching or scuffing. A widely accepted hardness classification for material in general is Mohs' scale, with ten degrees of hardness ranging from talc to diamond. On this scale any member will scratch all members lower down the order. Many methods of measuring hardness have been developed, but they are not all immediately applicable to glazes. For example, metals show plastic flow and glasses brittle fracture at room temperature, so that the mechanism of surface damage may be very different in the two cases. However, Roberts (1965) has used a diamond indentation method to study glaze hardness. He found that glaze composition had very little effect on hardness over a wide range of glaze types, most of which fell within the range 5-6 on Mohs' scale.

It is found in these studies that there are two clearly distinguishable types of damage. The first type is scratching, caused by contact with a sharp point or edge, such as a knife blade. There are two levels of scratching, as shown in Fig. 19.2. At the lower level, as shown in (a), the scratch is in the form of a thin line, or lines, which is scarcely visible to the naked eye. More severe scratching, clearly visible, is found under the microscope to have a herring bone pattern of minute cracks associated with the lines, as shown in (b). It is clear that there is a complicated pattern of stress at work here, including pressure, some frictional forces, and localized melting. As the point moves across the glaze surface it has a region of tension behind it, where cracking occurs.

The second type of damage is scuffing caused by the abrasion of one article on another, which occurs regularly; for example, in the stacking of plates. It is important to note here that any material will damage a material of the same hardness, however hard it is. This type of scuffing produces a gouging out of the glaze surface, as shown in (c), quite different in visual effect from scratching. It is impossible to avoid this kind of abrasion in service, though there have been many attempts to minimize it. In the glass industry, for example, coating the surface with a metal organic compound has been shown to be beneficial. Another idea is to have glazes of different hardness on the upper and lower faces of flatware, the damage certainly being reduced by this method. However, a solution that is practical for large-scale tableware production is not easy to find.

A special case of damage that merits a mention is metal or silver marking. Cutlery made of soft metal can sometimes leave a mark on the glaze surface. If there is no structural damage, this mark is easily removed, but if there is damage the metal may be embedded in such a way that it cannot be removed. Certain types of cutlery are more prone to produce this effect than others.

## 19.5 CLEANING

Of necessity, most ceramic articles, and more particularly tableware and sanitary ware, become contaminated in use, and regular cleaning is necessary to maintain

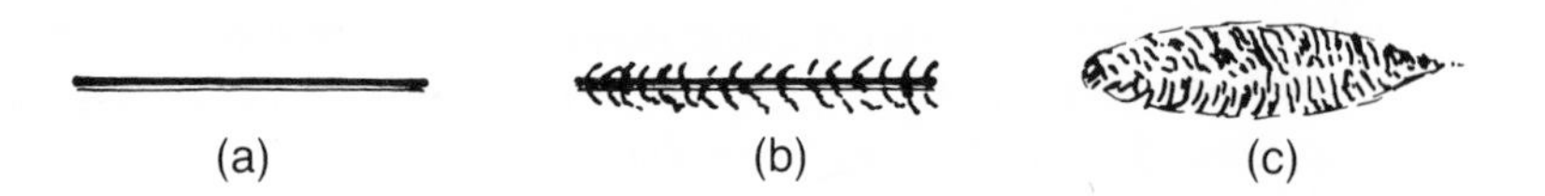


Fig. 19.2 – Different types of abrasion damage on a glaze surface.

gloss and general appearance. A word of warning seems appropriate concerning some cleaning agents that may cause damage, either by physical or chemical means. Some cleaners contain abrasives. If we look again at Mohs' scale, we find glazes with a value of 5-6, with quartz at 7, corundum at 9, and carborundum even harder. All these materials, then, will readily scratch glazes. Cleaning agents containing these abrasives should never be used on glaze surfaces.

Sanitary cleansers sometimes have a high sulphate content. When they are left in contact with glazes for a long period of time, as they often are in lavatory pans, for instance, the sulphuric acid formed in the wet conditions can attack the glaze surface.

In point of fact, neither of these aggresive types of cleaner is necessary. There are many harmless detergents that will maintain glaze surfaces in good condition, provided that they are used frequently.

Glaze surfaces are relatively easy to clean because they are easily wetted by water, having a very low angle of contact compared with the surface of metals, for example. That this is so can be demonstrated by examining the profiles of drops of water on various surfaces, as shown in Fig. 19.3. The angle of contact between the water and the surface increases markedly from glaze to plastic to metal. Problems of staining are nearly always due to contamination that has not been cleaned off, so that a hardened film or encrustation has been allowed to develop.

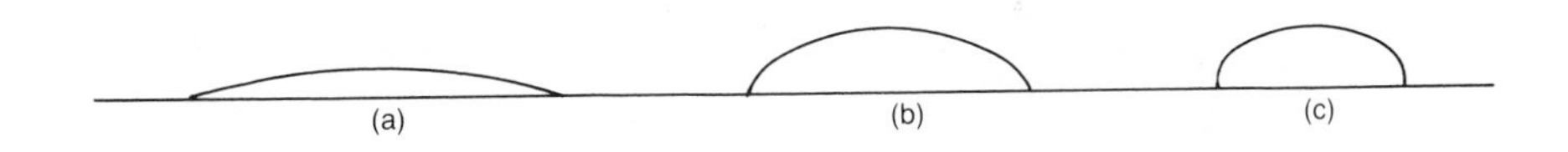


Fig. 19.3 – Profile of a drop of water on the surface of (a) glaze, (b) plastic, and (c) stainless steel.

## REFERENCES

Dinsdale, A. and Vaughan, F. (1962) *Tr. Brit. Cer. Soc.*, **61**, 1.

Franklin, C. E. L. and Tindall, J. A. (1959) *Tr. Brit. Cer. Soc.*, **58**, 589.

Harrison, R. and Dinsdale, A. (1964) *Tr. Brit. Cer. Soc.*, **63**, 63.

Moulson, A. J. and Roberts, J. P. (1960) *Tr. Brit. Cer. Soc.*, **59**, 388.

Norris, A. W., Vaughan, F., Harrison, R. and Seabridge, K. C. J. (1958) *International Ceramic Congress, VI,* 63.

Roberts, W. (1965) *Tr. Brit. Cer. Soc.*, **64**, 33.

Symposium on glazes. *Tr. & J. Brit. Cer. Soc.*, **79**, 2, xxxii.

Appendix 1

# Units and conversion factors

With few exceptions, SI units have been used throughout this book. However, much industrial practice is still based on more traditional units, such as CGS, or even occasionally FPS. A number of the more relevant conversion factors are given here.

| *Quantity* | *Unit* | *Conversion Factor* |
|---|---|---|
| Dimensional | | |
| Length | 1 metre m | = 39.37 in |
| | | = $10^6$ μm |
| | | = $10^{10}$ Ångstroms Å |
| Area | 1 $m^2$ | = $1.55 \times 10^3$ $in^2$ |
| Volume | 1 $m^3$ | = $6.102 \times 10^4$ $in^3$ |
| | | = 1760 pints |
| | | = 220 gallons |
| Mechanical | | |
| Mass | 1 kilogram kg | = $10^{-3}$ tonnes |
| | | = $9.842 \times 10^{-4}$ tons |
| | | = 2.205 lb |
| | | = 35.27 oz |
| Density | 1 kg $m^{-3}$ | = $10^{-3}$ g $ml^{-1}$ |
| | | = $6.243 \times 10^{-2}$ lb $ft^{-3}$ |
| | | = $2 \times 10^{-2}$ oz $pint^{-1}$ |

| | | |
|---|---|---|
| Force | 1 Newton N | = 0.2248 lbf |
| Stress | 1 $Nm^{-2}$ | = 1 Pascal Pa |
| | | = $1.450 \times 10^{-4}$ lbf $in^{-2}$ |
| Pressure | 1 atmosphere | = 101 325 $Nm^{-2}$ |
| | | = 1 013 250 dyn $cm^{-2}$ |
| | | = 760 mm Hg |
| | | = 14.696 lbf $in^{-2}$ |
| Energy | 1 Joule J | = 1 Nm |
| | | = $10^7$ erg |
| | | = 1 watt sec |
| | | = 0.737 ft lbf |
| Surface tension | 1 $Nm^{-1}$ | = $10^3$ dyn $cm^{-1}$ |
| Viscosity | 1 kg $m^{-1}$ $s^{-1}$ | = 10 poise |
| **Thermal** | | |
| Temperature | °K | = °C + 273 |
| Quantity of heat | 1J | = $2.389 \times 10^{-1}$ cal |
| | | = $9.48 \times 10^{-4}$ Btu |
| | | = $9.48 \times 10^{-9}$ therms |
| Atomic heat capacity | | = 5.96 cal gm $atom^{-1}$ |
| | | = 24 940 J kg $atom^{-1}$ |
| Specific heat capacity of water | | = $4.185 \times 10^3$ J $kg^{-1}$ $K^{-1}$ |
| | | = 1 cal $g^{-1}$ $°C^{-1}$ |
| Latent heat of water | | = $2.27 \times 10^6$ J $kg^{-1}$ |
| | | = 543 cal $g^{-1}$ |
| Thermal conductivity | 1 J $s^{-1}$ $K^{-1}$ $m^{-1}$ | = 1 W $K^{-1}$ $m^{-1}$ |
| | | = $2.389 \times 10^{-3}$ cal $s^{-1}$ $K^{-1}$ $cm^{-1}$ |

Appendix 2

# Standards, specifications and methods of test

Below is given a list of the more important standards, specifications and methods of test for pottery materials and products, published in Britain, United States and Germany.

BS = British Standards Institution.
ASTM = American Society for Testing and Materials.
ANSI = American National Standards Institute.
DIN = Deutsches Institut fur Normung.

**TABLEWARE**

| | |
|---|---|
| BS 4034. | The requirements for resistance to water absorption and crazing of vitrified hotelware. |
| BS 4860. | Permissible limits of metal release from glazed ceramic ware.<br>Part 1 Tableware.<br>Part 2 Cooking ware. |
| BS 5416. | Specification for water absorption and translucency of china or porcelain tableware. |
| ASTM C-738. | Standard test method for lead and cadmium extracted from glazed ceramic surfaces. |
| ASTM C-368. | Standard test method for impact resistance of ceramic tableware. |
| DIN 51032. | Limits for metal release. |

**TILES**

| | |
|---|---|
| BS 6431. | (Parts 1-21. covering a wide range of characteristics of wall and floor tiles.) |
| ANSI A137.1 | Specification for ceramic tile. |
| ASTM C-484 | Standard test method for thermal shock resistance of glazed ceramic tile. |
| ASTM C-499. | Standard test method for facial dimensions and thickness of flat, rectangular ceramic wall and floor tile. |
| ASTM C-502. | Standard test method for wedging of flat, rectangular ceramic wall and floor tile. |
| ASTM C-648. | Standard test method for breaking strength of ceramic tile. |
| ASTM C-1026. | Standard test method for measuring the resistance of ceramic tile to freeze-thaw cycling. |
| ASTM C-1027. | Standard test method for determining visible abrasion resistance of glazed ceramic tile. |
| DIN 18155. | Pars 1-4. Definitions, shapes and dimensions, quality and testing for ceramic tiles. |
| DIN 18156. | Ceramic split tiles. |

**SANITARY WARE**

| | |
|---|---|
| BS 3402. | Quality of vitreous china sanitary appliances. |
| ANSI A112.19.2 | Vitreous china plumbing fixtures. |

**ELECTRICAL PORCELAIN**

| | |
|---|---|
| BS 1598. | Ceramic insulating materials for general electrical purposes. |
| DIN 40685. | Ceramic insulating materials for electrical engineering; classification and technical data. |

**GENERAL**

Standard methods of test have been published by ASTM covering the following properties of ceramic materials:

| | |
|---|---|
| C-242. | Definitions of terms. |
| C-370. | Moisture expansion. |
| C-372 | Thermal expansion. |
| C-373. | Water absorption, bulk density, apparent porosity and specific gravity. |
| C-408. | Thermal conductivity. |
| C-424. | Crazing resistance. |
| C-554. | Thermal shock resistance. |
| C-584. | Gloss. |
| C-674. | Elastic properties. |
| C-773. | Crushing strength. |
| C-848. | Elastic properties by resonance. |
| C-949. | Porosity by dye penetration. |

Methods of test have also been issued by the Physical Testing Sub-committee of the British Ceramic Society covering water absorption, true and apparent porosity, crazing resistance, moisture expansion, thermal expansion, and particle size determination by Andreasen pipette and hydrometer.

EEC Directive 84/500 deals with metal release from ceramic articles.

# Name Index

# Subject Index